Modeling and Simulation
Challenges and Best Practices for Industry

T0175704

Modeling and Simulation
Challenges and Best Practices for Industry

Guillaume Dubois

CRC Press
Taylor & Francis Group
Boca Raton London New York

CRC Press is an imprint of the
Taylor & Francis Group, an **informa** business

This book was previously published in French as *La simulation numérique: Enjeux et bonnes pratiques pour l'industrie* (Numerical simulation: Challenges and best practices for industry) by Dunod, Malakoff, France.

CRC Press
Taylor & Francis Group
6000 Broken Sound Parkway NW, Suite 300
Boca Raton, FL 33487-2742

First issued in paperback 2021

© 2018 by Taylor & Francis Group, LLC
CRC Press is an imprint of Taylor & Francis Group, an Informa business

No claim to original U.S. Government works

ISBN-13: 978-0-367-78138-5 (pbk)
ISBN-13: 978-0-8153-7489-3 (hbk)

This book contains information obtained from authentic and highly regarded sources. Reasonable efforts have been made to publish reliable data and information, but the author and publisher cannot assume responsibility for the validity of all materials or the consequences of their use. The authors and publishers have attempted to trace the copyright holders of all material reproduced in this publication and apologize to copyright holders if permission to publish in this form has not been obtained. If any copyright material has not been acknowledged please write and let us know so we may rectify in any future reprint.

Except as permitted under U.S. Copyright Law, no part of this book may be reprinted, reproduced, transmitted, or utilized in any form by any electronic, mechanical, or other means, now known or hereafter invented, including photocopying, microfilming, and recording, or in any information storage or retrieval system, without written permission from the publishers.

For permission to photocopy or use material electronically from this work, please access www.copyright.com (http://www.copyright.com/) or contact the Copyright Clearance Center, Inc. (CCC), 222 Rosewood Drive, Danvers, MA 01923, 978-750-8400. CCC is a not-for-profit organization that provides licenses and registration for a variety of users. For organizations that have been granted a photocopy license by the CCC, a separate system of payment has been arranged.

Trademark Notice: Product or corporate names may be trademarks or registered trademarks, and are used only for identification and explanation without intent to infringe.

Library of Congress Cataloging-in-Publication Data

Names: Dubois, Guillaume, author.
Title: Modeling and simulation : challenges and best practices for industry / Guillaume Dubois.
Description: Boca Raton : Taylor & Francis, a CRC title, part of the Taylor & Francis imprint, a member of the Taylor & Francis Group, the academic division of T&F Informa, plc, 2018. | Includes bibliographical references.
Identifiers: LCCN 2017054207| ISBN 9780815374893 (hardback : acid-free paper) | ISBN 9781351241137 (ebook)
Subjects: LCSH: Simulation methods.
Classification: LCC T57.62 .D83 2018 | DDC 620.001/1--dc23
LC record available at https://lccn.loc.gov/2017054207

Visit the Taylor & Francis Web site at
http://www.taylorandfrancis.com

and the CRC Press Web site at
http://www.crcpress.com

"Be the change that you wish to see in this world."

—Gandhi

Foreword

Numerical modeling and simulation have revolutionized the industry over the past few decades. It has allowed us to design more complex systems and predict behaviors without field testing. Today's trucks, cars, and planes are higher performing and produced faster at a lower cost. However, numerical simulation expansion and mutations are still evolving at pace and have a long way to go. Great opportunities for numerous industries are still to come.

My teams and I often have to address challenging questions that require the use of numerical simulation. To go further, the scope of simulation, its limitations, the expected impacts and benefits must be assessed. These are questions that are top of the mind for any company that wants to remain competitive and relevant in the future.

This book is accessible and holistic enough to be useful to different categories of readers, whether they are project leaders, engineers, or students. It gives a good understanding of what simulation is and when it really should be used. It further provides both technical and organizational best practices to implement it. This book will be a strong contribution to the expansion and adoption of simulation in the industry.

Martin Lundstedt,
President and CEO of Volvo Group

Special Thanks

I thank, for proofreading or making detailed comments on this or that, in alphabetical order: Alexis Beauvillain, Claire Casas, Jyster Crauser-Delbourg, Laurent Di Valentin, Sébastien Dubois, Antoine Ferret, Manuel Fontanier, Florent Fossé, Étienne Fradet, Éric Le Dantec, Ikramjit Narang, Éric Noirtat, Hugues-Loup Robedat, Éric Suty, Manuel Tancrez; Éric Tomieto, Benoît Trémeau, Yang Xu, and Alexandre Zann.

To Taylor & Francis, especially Cindy Renee Carelli, executive editor, and Renee Nakash, editorial assistant, for their professionalism and the quality of their work.

To Dunod for allowing me to publish this book abroad.

To Martin Lundstedt, for his Foreword and being a great leader.

To Dorothée Dorwood and Olivier Franceschi, the Traductorz, for the quality of their translation and their great conviviality.

To Benoît Parmentier, whom fate took away violently and whom I wish could have proofread this book, for training me when I was starting out.

To my parents, Chantal and Xavier, for their proofreading, their encouragements, and their eternal support.

To my family, especially Cécile, Sébastien, Claire, Youri, Lucas, Julien, Jeanine, Gérard, Martine, Séverine, Ivan, Geneviève, and Michel, for encouraging me and bringing me happy moments.

To Violette, for being inspiring.

To the Ethiopian farmers from 2000 years ago who discovered coffee and enabled me to write this book within a few weeks, by night.

Finally, to Doug, the fictional engineer we follow in this book, who, although he doesn't exist, helped me a lot in my writing.

Contents

About the Author

Guillaume Dubois works at McKinsey & Company and serves clients in various types of industries.

Guillaume graduated as an engineer from École Centrale Paris, France. He started his career at Pratt & Whitney Canada, developing thermal models for airplane engines and at Enertime, developing financial models for biomass plants. He then joined PSA Peugeot Citroën, where he created and grew new modeling activities, teams, and processes across R&D departments to improve energy efficiency of cars.

He has progressively held the different roles discussed in this book: modeling engineer—developing models, team manager—managing these engineers, project manager—being a client of these teams.

Guillaume authored *La simulation numérique: Enjeux et bonnes pratiques pour l'industrie* (Numerical simulation: Challenges and best practices for industry), published by Dunod, Malakoff, France, in 2016. *Modeling and Simulation: Challenges and Best Practices for Industry* is the English version.

This book has been translated by Dorothée Dorwood and Olivier Franceschi.

Introduction

Why This Book?

A few decades ago, numerical simulation introduced a significant disruption among the major sectors of industry. It revolutionized the realm of possibility and brought more value than the usual operational tools. It then expanded progressively. In the early twenty-first century, numerical simulation expansion and mutations are far from being over. Great changes are still ahead of us.

However, the expansion of numerical simulation remains obviously slow. Only a few industry leaders manage to comprehend its full potential. We'll see that this slowness may be partly explained by poor knowledge sharing. This sluggishness affects its expansion in industry and results in a loss of potential return for our society.

The usual questions, which don't often receive satisfying answers, are the following:

1. What is numerical simulation and what is its intrinsic value?
2. Why expand it further in industry? Under what conditions? How much trust can we allow?
3. How can it be expanded in industry?
4. What are the pitfalls and the technical and organizational best practices?
5. What leads are there for the future? How can we be a driving force?

The purpose of this work is to bring the clearest answers to these questions. Our goal here is to contribute to knowledge sharing in order to open pathways to the expansion of numerical simulation and maximize the creation of value for our society.

Observation: The objective of this work is not to explain in detail digital methods and techniques adapted to specific fields. Many works exist about these topics (including *Pratique de la simulation numérique*, Dunod).

Who Is This Book For? (To Be Read First If You Are in a Hurry)

This work targets three types of people:

- *Managers and project managers*: In order to gain perspective on the subject of numerical simulation, the conditions of successful planning, the limits, and the impacts on management, first read Chapters 1, 3, 5, and 6.

- *Modeling and simulation engineers (in every sense, including experts)*: In order to contribute from the inside to the expansion of simulation, still gaining perspective on the impacts, first read Chapters 1, 3, 4, and 6.

- *Students and teachers*: In order to prepare the next generation, who will contribute to simulation in industry, first read Chapters 1, 2, 3, 4, and 6.

What Industries Is This Work Concerned With?

Numerical simulation is used in an increasing number of sectors. Nevertheless, the present work focuses on its use in industry, particularly the aircraft, energy, car, space, transport, agribusiness, and chemistry industries.

1

What Is Numerical Simulation?

A model is a testable representation of a system. Numerical simulation consists of translating a mathematical model into numerical language in order to perform complex calculations more easily.

Discerning readers may skip to the next chapter.

1.1 What Is a Model?

A model of a *system* is an *experienceable representation*.

Let's begin with defining what a system is. It is an object (real or abstract) or a set of objects that we are willing to study. It is a very broad concept: for instance, a system may be a plane, an engine, or even a screw. Note that it may contain subsystems that are systems as well. So, in industry, teams work on systems on a daily basis (these systems will be products or parts of products marketed by the company), either to conceive, develop, or produce them.

There are many possible representations of a single system. In this work, we will deal with a mathematical representation; that is to say, the representation aims at describing its behavior through equations. Those equations often come from known laws of physics (such as the forces of gravity or the fundamental principles of dynamics). The representation of a system may be understood as a rationalization of reality or a vision of a system through one angle.

In industry, a wide range of systems are constantly tested through live experiments. For example, we are able to test an aircraft engine to assess the temperatures its compartments can reach or to test a screw to determine what mechanical stresses it can resist. Also, a system representation can be tested. This experiment will consequently be virtual.

Let's illustrate this concept with an example, represented by the figure below, assuming that our goal is to assess the duration of an ice cube melting in a glass of water (Figure 1.1).

The system studied here is an ice cube—that is to say, the volume of water initially contained in it. The rest of the water filling the glass will be the

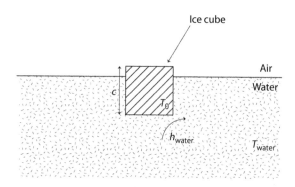

FIGURE 1.1
The system studied: an ice cube floating in a glass of water.

environment of the system. We will build a representation of the ice cube, making it simpler.

We consider that the ice cube exchanges, by convection, with the rest of the water contained in the glass. Using thermodynamics, we can assume that the exchanged heating power will be

$$P_{\text{exchanged}}\left(t\right)=5h_{\text{water}}\left(T_{\text{water}}-T_0\right)c^2\left(t\right)$$

Where:

t is time
h_{water} is the heat convection coefficient between the water and the ice cube
T_{water} is the water temperature
T_0 is the temperature of the ice cube
$c(t)$ is the width of the cube, still frozen

Our hypothesis is that the heat exchanges with the air, at the upper side of the ice cube, are insignificant compared with those of the water, and we are simplifying the problem by also assuming all the other sides are fully in contact with the water.

Heat energy contained in the ice cube can be approximated as

$$E_{\text{cube}}\left(t\right)=E_0-L_{\text{water fusion}}\,\rho_{\text{ice}}\,c^3\left(t\right)$$

Where:

E_0 is the ice cube's energy once it has melted
$L_{\text{water fusion}}$ is the mass enthalpy of the state change from a liquid to a solid
            state

Usual physics laws allow us to predict the total exchanged power equal to the variation of the system's energy. Thus, we can write

$$\begin{cases} P_{\text{exchanged}}(t) = 5h_{\text{water}}(T_{\text{water}} - T_0)c^2(t) \\ E_{\text{cube}}(t) = E_0 - L_{\text{water fusion}}\,\rho_{\text{ice}}\,c^3(t) \\ \dfrac{\partial E_{\text{cube}}(t)}{\partial(t)} = P_{\text{exchanged}}(t) \\ c(t_{\text{final}}) = 0 \end{cases}$$

FIGURE 1.2
Mathematical model vs. reality.

This set of equations

- Is a system representation (the ice cube)
- Is testable (It is a set of four equations with four unknowns and a single solution we can determine.) (Figure 1.2)

This is a model!

To thoroughly illustrate our example, we can test this model on a specific situation. The model is solved analytically, and the solution of this set of equations is (calculation details are skipped)

$$t_{\text{final}} = \frac{3L_{\text{water fusion}}\,\rho_{\text{ice}}\,c(t=0)}{5h_{\text{water}}(T_{\text{water}} - T_0)}$$

Where t_{final} is the time the ice cube takes to melt entirely.

Let's detail the situation we want to test.

$$\begin{cases} c(t=0) = 2 \text{ cm} \\ T_{\text{water}} = 20^\circ\text{C} \\ T_0 = 0^\circ\text{C} \\ h_{\text{water}} = 250 \text{ W/K/m}^2 \\ L_{\text{water fusion}} = 334 \text{ kJ/kg} \\ \rho_{ice} = 917 \text{ kg/m}^3 \end{cases}$$

The result we get for this test is $t_{\text{final}} \approx 12$ mn.

We have performed the live experiment for you, reader, and the time found was 9 minutes, which means that our model has a +33% gap with reality. Nothing surprising here, as our modeling is very simplified and we have eluded several factors.

A significant consequence of this definition: there is always an infinite range of models for a single system. Going back to the example of the ice cube, the reader will notice that we could have complicated the system representation (we would theoretically get a more accurate time for the duration of the ice cube melting).

We could have taken into account the ice cube's surface reduction and its shape-changing effects on the power exchange between the air and the cube, or the radiative occurrences between the ice cube and the other surrounding elements, which can't be ignored. We made a *choice* as to the level of reality simplification to build this model. This choice must depend on many factors, which will be covered in Chapter 3.

1.2 What Is a Simulation?

A *simulation* is the action of performing a test on a model.

Theoretically speaking, the concepts of model and simulation are different, as a model is a tool, whereas a simulation is the action of using that tool. Nevertheless, these notions are connected and in practice are often used interchangeably. The result of a simulation is also frequently called a simulation.

1.3 What Are Modeling and Numerical Simulations?

A *numerical model* is a model implemented into a numerical tool.

In practice, it is about translating the mathematical model into a numerical language (informatics) in order to test it on a numerical tool (computer).

Why use numerical tools?

Digital technology is used because we don't know how to get results on a complex model without using a digital tool. Today, any computer will calculate faster than a human. Whenever the calculations become numerous or complex, they become theoretically unachievable without a digital tool. Indeed, it would require too much time to get the result.

Let's illustrate this by going back to the last example of an ice cube melting. We took a very simple case, where we could solve the model without needing a digital tool. Let's make the situation a little more complex, assuming this time that a small heated steel solid was placed at the surface of the ice cube (Figure 1.3).

This heated steel solid will increase the melting speed of the ice cube. We can go back to the previous equations, considering that an additional exchange is taken into account: the heat exchange with the steel solid. Let's assume that this extra power is $P_{\text{solid exchange}}(t) = \varphi_{\text{solid}}$, where φ_{solid} is a constant equaling 1 W. The model becomes

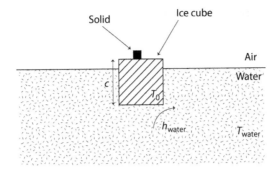

FIGURE 1.3
The new system studied: the ice cube is now touching a solid made of steel.

$$\begin{cases} P_{\text{exchanged}}(t) = 5h_{\text{water}}(T_{\text{water}} - T_0)c^2(t) + \varphi_{\text{solid}} \\[2mm] E_{\text{cube}}(t) = E_0 - L_{\text{water fusion}}\rho_{\text{ice}}\,c^3(t) \\[2mm] \dfrac{\partial E_{\text{cube}}(t)}{\partial t} = P_{\text{exchanged}}(t) \\[2mm] c(t_{\text{final}}) = 0 \end{cases}$$

or

$$\begin{cases} \dfrac{\partial c(t)}{\partial t} = -\left[A + B\dfrac{1}{c^2(t)}\right] \\[3mm] A = \dfrac{5h_{\text{water}}(T_{\text{water}} - T_0)}{3L_{\text{water fusion}}\rho_{\text{ice}}} \\[3mm] B = \dfrac{\varphi_{\text{solid}}}{3L_{\text{water fusion}}\rho_{\text{ice}}} \\[3mm] c(t_{\text{final}}) = 0 \end{cases}$$

This very simple addition (the heated solid) complicates the situation. Here, the equation isn't analytically solvable (without using inverse trigonometric functions; that would also require the digital tool).

We have to use numerical methods to solve the equation. For instance, we can reject time, considering an infinitesimal time frame dt (we will take here $dt = 0.1$ s).

$$\begin{cases} \dfrac{c(t+dt) - c(t)}{dt} = -\left[A + B\dfrac{1}{c^2(t)}\right] \\[3mm] c(t_{\text{final}}) = 0 \end{cases}$$

We are thereby able to calculate c step by step, until the value reaches 0.

$$\begin{cases} c(t=0)=c0 \\[2mm] c(dt)=c(t=0)-dt\left[A+B\dfrac{1}{c^2(t=0)}\right] \\[4mm] c(2*dt)=c(t=dt)-dt\left[A+B\dfrac{1}{c^2(t=dt)}\right] \\[4mm] c(3*dt)=c(t=2*dt)-dt\left[A+B\dfrac{1}{c^2(t=2*dt)}\right] \\[2mm] \cdots \\[2mm] c(n*dt)=c(t=(n-1)*dt)-dt\left[A+B\dfrac{1}{c^2(t=(n-1)*dt)}\right] \\[4mm] c(t_{\text{final}})=0 \end{cases}$$

In this way, there are n calculations to operate to be able to find the final value (n being the number of iterations to perform before stopping); that is to say, before $c(n*dt)$ becomes negative. We'll see that $t_{\text{final}} \approx 455$ s ≈ 8 mn, where $dt = 0.1$ s; so 4550 calculations must be performed to find the answer!

This is the moment the digital tool unveils its usefulness. It is able to perform these 4550 calculations in a split second, whereas a human being would have needed several hours. A computer just needs these previous equations translated into a digital language to solve them (see the numerical model in Figure 1.4).

By using a computer and the appropriate software, these calculations are performed in a split second, and the solution we find is $t_{\text{final}} \approx 455$ s ≈ 8 mn.

Thus, in 99% of the situations encountered in industry, mathematical models are converted into numerical models, in order to solve them with digital tools.

As explained in the previous paragraph, a *numerical simulation* is the action of performing a test with a numerical model.

1.4 What Is a State Representation?

We briefly mentioned this notion because it organizes the model and will be useful later. A model can take the following form (state representation):

```
T_water = 273.15 + 20
T_0 = 273.15 + 0
h_water = 250
L_water_fusion = 334 *10^3
Rho_ice = 917
Phi_solid = 1


A = (5 * h_water (T_water - T_0))/
    (3 * L_water_fusion * Rho_ice)
B = Phi_solid/(3 * L_water_fusion * Rho_ice)

c = 0.02
t = 0
Delta_t = 0.1

While (c > 0)
    c = c - Delta_t * (A + B/(c^2))
    t = t + Delta_t
End


Print t
```

$$\begin{cases} \dfrac{\partial c(t)}{\partial t} = -\left[A + B \dfrac{1}{c^2(t)} \right] \\[2mm] A = \dfrac{5h_{\text{water}}\ (T_{\text{water}} - T_0)}{3L_{\text{water fusion}}\ \rho_{\text{ice}}} \\[2mm] B = \dfrac{\varphi_{\text{solid}}}{3L_{\text{water fusion}}\ \rho_{\text{ice}}} \\[2mm] c(t_{\text{final}}) = 0 \end{cases}$$

FIGURE 1.4
Numerical mode, mathematical model, and reality.

$$\begin{cases} \dfrac{d}{dt} x = f(x, u, \theta, t) \\[2mm] y = h(x, u, \theta, t) \end{cases}$$

Where:

- x is the state vector, including variables selected to define the system
- y is the output vector, including the variables to be observed
- u is the input vector, including the variables defining the system request
- θ is the setting vector, including the setting variables of the system

We can illustrate this with the previous example:

$$\begin{cases} \dfrac{\partial c(t)}{\partial t} = -\left[\dfrac{5h_{\text{water}}(T_{\text{water}} - T_0)}{3L_{\text{water fusion}}\ \rho_{\text{ice}}} + \dfrac{\varphi_{\text{solid}}}{3L_{\text{water fusion}}\ \rho_{\text{ice}}} \dfrac{1}{c^2(t)} \right] \\[2mm] y = c(t) \end{cases}$$

Where:

$x = c(t)$

$y = c(t)$

$u = [T_{\text{water}}\ T_0\ \varphi_{\text{solid}}]$

$\theta = [h_{\text{water}}\ L_{\text{water fusion}}\ \rho_{\text{ice}}]$

The model being simple, we only have a single state variable, and the output variable is equal to the state variable.

1.5 What Is the Value of Numerical Simulation?

We stated that a model is a testable representation of a system, which leads to two inherent values:

1. *A model is testable*: It is possible to predict, without live experiments, the behavior of the system in a specific situation.
2. *A model is a representation*: It is necessary to comprehend the system to represent it, and the model itself includes this knowledge.

These two core values, combined with the power of the tools of digital technology, generate huge potential to support industrial companies. More details will be covered in Chapter 3.

2

A Bit of History

Simulation has existed for millennia; it became numerical and started its escalation in the 1940s thanks to computers. Its history is tied in but not restricted to theirs, considering the many internal innovations that have occurred. Numerical simulation kept growing once it arose and went through a strong acceleration in the 1980s.

2.1 Before 1940

Before 1940, we can say that modeling and numerical simulation didn't exist. Two points have to be highlighted.

1. *Modeling (non-numerical) has existed for millennia*: We have established that modeling consists in representing a system to study it. So, this doesn't necessarily require tools. For instance, back in the seventeenth century, using Newton's laws of gravity to represent the falling of an apple and its speed already involved modeling. Nevertheless, this type of modeling didn't use digital hardware.

2. *Calculation tools existed long before 1940*: Before the emergence of computers as we know them, there were many calculation tools. Take Blaise Pascal's mechanical calculating machine (Figure 2.1), for example, which allowed one to perform calculations with large numbers as early as the seventeenth century. While the machine itself was a commercial flop, it can be considered a pioneer of computers; however, the tools available at the time were not able to be automated, and their performance was clearly nothing compared with that of the first computers.

The emergence of the first computers was necessary for the rise of numerical simulation.

FIGURE 2.1
Blaise Pascal's calculating machine. (© 2005 David Monniaux.)

2.2 From 1940 to 1960: The First Steps of Numerical Simulation

Over the 1940s and the 1950s, numerical simulation made its first steps, although it hadn't really reached industry yet.

The 1940s marked the beginning of computer science. The first programmable calculators were built a few years before, and the Second World War stimulated research. Toward the end of the war, the creation of the Electronic Numerical Integrator and Computer (ENIAC; Figure 2.2) went down in history. It was the very first fully digital computer able to be programmed in order to solve calculatory problems. Admittedly, it cost half a million dollars, weighed 30 tons, occupied 160 m² on the ground, and consumed 150 kW of power. Nevertheless, the ENIAC allowed one to calculate much faster than humans by processing nearly 100,000 elementary additions per second (to provide an order of magnitude, current computers are about one million times more powerful).

The first real numerical simulation works are often associated with the Manhattan Project. This research project, carried out by the United States in partnership with the United Kingdom and Canada, led to the production

FIGURE 2.2
The ENIAC, the first fully digital computer. (From: Kempf, K., "U.S. Army Photo"; Kempf, K., "Historical Monograph: Electronic Computers within the Ordnance Corps".)

FIGURE 2.3
John von Neumann, one of the major pioneers of numerical simulation in the mid-twentieth century. (© 2016 Los Alamos National Security, LLC.)

of the first atomic bomb during the Second World War. John von Neumann and his coworkers then used the first computers to help this project. The first models being simulated represented the nuclear detonation process. The algorithms used the *Monte Carlo method*, one of the stochastic methods we'll cover in Chapter 3 (Figure 2.3).

The methods of numerical simulation then started broadening. Keith Douglas Tocher made a mark at the time, especially by setting up the *three-phase method*, which is still used nowadays for discrete simulations. The first phase consists in going to the next-time step. The second phase consists in calculating all events that are able to be calculated at that instant. The last one then consists in calculating all events occurring, providing specific factors that might have been affected by the calculations in the second phase. These three phases are repeated as many times as necessary. Tocher used this method to develop a model allowing one to simulate the operations of a production factory, integrating the flows of raw materials and the different stocks. He wrote *The Art of Simulation* in 1963, which was the very first book about numerical simulation.

Numerical simulation in industry was still very marginal (Tocher's works in the late 1950s and early 1960s are some of the rare exceptions). Its purposes were, at the time, mainly scientific.

2.3 From 1960 to 1980: The Evolution of Numerical Simulation

Over the 1960s and the 1970s, numerical simulation started developing and reaching industry. Two different evolutions that both occurred at this time have to be distinguished: those of computer science and numerical simulation.

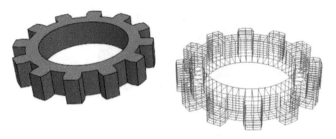

FIGURE 2.4
With the rise of CAD, industrial drawings were then made on computers.

Indeed, informatics boomed dramatically with the massive development of computers during this period. As early as 1963, computers were using circuit boards, making them more compact. Many more innovations were made, and new manufacturers entered the market. Computers became more powerful, their prices lowered, and their number increased. What was thousands of computers quickly became hundreds of thousands.

Even if it is undeniable that simulation benefited from the progress of numerical tools, numerical simulation specific to innovations also changed the game.

- *Simulation software*: The first software dedicated to numerical simulation then appeared, easing modeling tasks. It quickly multiplied or was renewed with new features. For the first time, simulation became affordable to engineers who weren't well versed in computer science.

- *Computer-aided design (CAD)*: CAD (Figure 2.4) appeared in the late 1950s and quickly spread, allowing engineers to visualize in 2D, then in 3D, parts that were then virtual. Several aircraft and car industries adopted this revolutionary method. Converting drawn designs into numerical designs was hard work, but it allowed one to detect many human mistakes, and digital technology enabled one to avoid them from then on (inaccurate scaling, for instance).

  The improved robustness of digital technology was a forceful argument to back its expansion.

- *Winter simulation conference*: In 1967, the first world conference focused on simulation was held. Its purpose was to introduce new insights within the field, gathering researchers, engineers, and commercial enterprises from all sectors, including industry, government, the military, and education. These annual conferences are still held today, and the content of the presentations is copyright free. Among the main topics being discussed, it is relevant to notice that by the late 1970s, current topics were already being covered. The challenge of mastering simulation tools, debugging code, or convincing managers with calculation results were often discussed. Despite many evolutions, many issues related to numerical simulation are still the same today (we'll cover them in the upcoming chapters).

Consequently, it is during this period that industries really started benefiting from numerical simulation, albeit moderately.

2.4 From 1980 to 1995: The Revolution of Numerical Simulation

The 1980s and the early 1990s were a turning point in the numerical simulation expansion in industry. This period is certainly when numerical simulation expanded the most, although the previous and next periods experienced large growth.

In the late 1970s and the early 1980s, a major disruption occurred in the field of computer science: the rise of the personal computer (Figure 2.5). Because of both ruthless competition and continuous innovations in the field, manufacturers were able to market computers the middle class could afford. Computers became more accessible in both private and professional environments. There was no longer a need for specific skills to use them. Thus, in the late 1980s, most companies were computer equipped. Now that computers were more accessible, numerical simulation could then expand massively in industry.

- *Modeling on new topics*: Over this period of time, simulation expanded a lot, entering more and more companies in diverse fields, and dealing with more and more various issues.
- *Software consolidation*: Some software programs stood out and slowly became the must-haves of their own sectors. So, many of the programs used today were edited at this time. They were more

FIGURE 2.5
One of the first personal computers of the 1980s. (© Bill Bertram 2006, CC-BY-2.5.)

ergonomic than the previous ones and helped the modeler to focus on modeling, making numerical issues almost nonexistent.

- *CAD's long-term establishment*: The use of CAD became progressively inevitable. Numerical models contained geometrical representations of the studied system (including products in the conception process) and enabled one to analyze, control, and simulate some behaviors. Numerical models became the unique toolkit product for companies.

  Thanks to software improvements, CAD works became less and less time-consuming (it was then possible to define configured scaling and modify it afterward).

- *Visual animations*: As simulation results were previously represented on boards or as graphics, many programs were developed allowing one to dynamically visualize the results of the simulation. This enabled one to improve the understanding and communication of results.

2.5 From 1995 to 2015: The Spread of Numerical Simulation

Several further innovations have marked the last 20 years, helping numerical simulation to keep on spreading to more industrial companies.

Regarding computer innovations, besides the continuous progress of computers, a major change occurred in the 1990s: The Internet. In 1993, the European Council for Nuclear Research (ECNR) put the very first website online, allowing most of the known content created on the Internet to use the World Wide Web protocol. This worldwide network became accessible to anyone and revolutionized both personal and professional lifestyles. Users were connected on a worldwide scale and could easily share documents and information (Figure 2.6).

FIGURE 2.6
The Internet has enabled users to connect worldwide since the 1990s.

Several evolutions occurred in numerical simulation, making it more profitable for industry.

- *Internet impact*: In the end, its significance on numerical simulation remained moderate, despite major lifestyle changes. Two impacts on numerical simulation may be mentioned. First, the Internet—and more widely, digital networks—had a direct impact. For example, it enabled distributed simulations, which consisted in performing several simulations on remote computer(s). One of the conveniences was the saving of calculation time (by performing the calculation on several powerful computers). This technique had yet to mature, although it had already been used and benefited various domains. Some issues still needed to be adjusted: the coding to distribute calculations, the setting up of networks, and the management of license amounts with the editors. The Internet and networks also had an indirect impact that was important. Thanks to the improvement of exchanges brought about by this innovation, communicating, spreading knowledge, and sharing models became easier; all of this had a positive impact on the efficiency of numerical simulation work.

- *Modeling on new thematics*: Again, modeling kept on developing new thematics. For instance, many optimization works were set up thanks to modeling. Indeed, many optimization issues came up in industry. They were not analytically solvable, and the simulations allowed one to identify solutions. Thus, software gained more and more additional content/packages to assist optimization works. Also, with its democratization, simulation services spread more and more, and on less technical topics.

- *Virtual reality*: Virtual reality started entering industry in the late 1990s. It consisted in simulating and revealing a system, in the most realistic way (3D), so that the user might observe or even interact with it. The applications were numerous, and two of them will be covered here. First, during the car design process, having the first physical prototype helps to determine whether the style of the car is satisfying. This is easier to evaluate than with 2D plans. Also, during the design process of a production line, it can be difficult to know whether the manipulations the operators will have to do are feasible or not, and whether ergonomic issues will come up. Instead of building a prototype production line, it was possible to model it and check in virtual reality whether an operator could perform the expected operations. In this way, virtual reality has expanded over the last 20 years in industry, but maturity is far from being reached, the realism of the means being used are still rather imperfect and the cost is often unacceptable.

- *Compatibility of software*: In the early 1990s, simulation software programs were mostly not designed to work together. This limited the opportunities offered by simulation, making the creation of extensive models more complex or, at least, requiring different abilities. Over the last two decades, programmers have created open pathways between types of software to lift these barriers. For example, it is common to use a specific CAD software to design shades and another to perform physical behavior studies as the connection is now possible. However, as with the two previous points, while huge progress has been made, these connections have not yet matured.

2.6 Three Lessons from History

The history of numerical simulation we have just covered allows us to highlight three points that will be useful later:

1. *The duration of this history is on a human scale*: We just mentioned that numerical simulation entered industry only 60 years ago and reached bigger proportions 40 years ago. There was also a discrepancy before it reached education. So, numerical simulation is "recent" on a human scale. This is why many industry actors are not well versed in simulation and keep seeing it as a "new" activity.

2. *Numerical simulation is always renewed*: It didn't follow the three common steps, which are start, growth, and maturity. Indeed, a large number of innovations have appeared during its history, whether internally or externally, so much so that it has been continuously renewed and has never reached maturity (besides a few specific fields of use). So, as this step hasn't yet been reached, there are large disparities in industry today: some industry stakeholders have become more advanced than others by integrating the emerging innovations in turn, more or less quickly.

3. *The history of numerical simulation is still being written today*: The greatest changes are still occurring. First, we've seen that yesterday's innovations are not yet perfectly distributed. And as we'll cover in Chapter 6, many innovations are still ahead of us, so the evolution of numerical simulation is not complete.

We just mentioned that the history of numerical simulation is far from being over. Now it is time to see, from an industry perspective, how profitable investing in numerical simulation can be.

3

Numerical Simulation in Industry: Why?

In what cases would one use numerical simulation? When is it a profitable investment?

The decision to use this technique has to be taken in a rational way, with an unbiased analysis of its outputs, limits, and costs.

3.1 Why Expand Simulation?

3.1.1 A Predictive Tool...

As shown in the introduction, simulation is a tool that allows one to forecast the behavior of a system, which is a crucial activity for industrial companies interested in the present publication.

For instance, to design a machine, measurement options must be chosen among several hypotheses. To do so, the results have to be predicted with simulation and the best will be selected.

This need for prediction is obviously not new. The designers of the very first steam engines also had to calculate the dimensions of their systems. How did they do it? They went through three levels of solutions. They used their own expert knowledge to draw the outlines of their machines, they built prototypes to adjust their choices, and they rejected others they couldn't test. They didn't wait for the emergence of simulation to make their predictions and build their machines!

So, there are three common answers to that need for prediction (or three alternatives to numerical simulation):

1. *Expertise* consists in using past experience and knowledge of the laws of physics. We also add non-numerical models here (as seen in the introduction, the purpose is to simplify the system and translate it into mathematical equations).

2. *Testing* is about building the system (potentially a prototype, a simplified version) and carrying out tests.

3. *Rejection* is ... a nonanswer. It is about deciding not to predict the behavior of the system (e.g., aborting the project or carrying it on without measuring the impact of the choices).

Nowadays, any industrial company should use these three levels to answer its need for predictions. Our goal is not to assert that these three methods are obsolete but to introduce a "new" (see Chapter 2 for a historical background) forecasting tool, numerical simulation, which has been spreading over recent decades. As we will see, the latter is more relevant in specific cases than the other three. This new option was the game changer that broke the balance in use.

Our first issue—"Why expand simulation?"—remains relevant but becomes "Why expand that new predictive tool instead of the three others already in use?" This is the purpose of the next section.

3.1.2 ...That Is Profitable

Why expand simulation rather than the alternatives in use? Because it is a profitable investment!

All the reasons for the development of simulation have financial ambitions. This may appear simplistic, or even provocative, and we'll explain it. We adopt this angle for three reasons.

First, this work deals with the contribution of simulation to industries that, for most of them, are under financial pressure. We won't argue that simulation may become art or that it may still be deployed in the name of artistry or science if it's not relevant.

Then, taking the profit angle doesn't mean ignoring the other important outcomes of simulation. For instance, we will not skip long-term contributions such as the ability to innovate more thanks to digital simulation; they will also support long-term profits.

Furthermore, the contributions detailed in this book will also be relevant for nonprofit organizations. Indeed, any company has efficiency objectives, and the analogy can be easily applied to financial efficiency.

Eventually, and above all, we will take the financial angle, which is potentially provocative, as the advantage of using numerical simulation or not often lacks objectivity.

- Some say, sometimes rightly, that digital simulation is an obsession for tech-friendly employees who entertain themselves by designing worthless models.
- Some others judge, sometimes rightly, that management lacks perspective and has very short-term vision, making the wrong decision not to invest in simulation even though it would be profitable.

We're trying here to make the debate unbiased and factual, to enlighten the decision of investing in simulation, instead of being the result of a misinformed hunch.

We will see in this chapter that simulation can be seen as an investment with:

1. Short- and long-term profit creation by
   - Increasing company incomes (e.g., decreasing *time to market*)
   - Reducing company expenses (e.g., decreasing the amount of physical tests)

     We'll detail this in Section 3.2.
2. Investment cost as expanding simulation means fees. We'll detail this in Section 3.3.

The decision to expand simulation or not can then be made based on financial considerations: it is relevant when profit balances its investment (or when return on investment exceeds a certain amount). We'll detail this topic in Section 3.4.

3.2 Contributions of Simulation

In this section, we will detail how profitable modeling can be to industry. Understanding how it leads to returns is crucial in order to calculate it and, consequently, to invest in digital simulation and maximize these benefits. The eight levers will be presented next.

3.2.1 The Modeling Iceberg

Before introducing the eight levers, let's briefly explain this notion of the modeling iceberg. Simulation benefits are generally unknown. Among the eight main benefits, only two are often quoted (cutting the costs of live experiments and decreasing the time to market). Plus, these two benefits are rarely the most decisive.

The reason why we mention this concept is mainly to warn the reader that benefits are not necessarily where we would expect them. This way, we should be neither too pessimistic by neglecting the benefits of the immersed part of the iceberg nor too optimistic by overestimating the benefits of the emerged part (Figure 3.1).

3.2.2 Eight Levers of Value Creation

We're going to point out the two types of benefits of numerical simulation:

1. *Intrinsic benefits*: These are the direct benefits that we started to introduce earlier, where simulation consists in testing a representation.

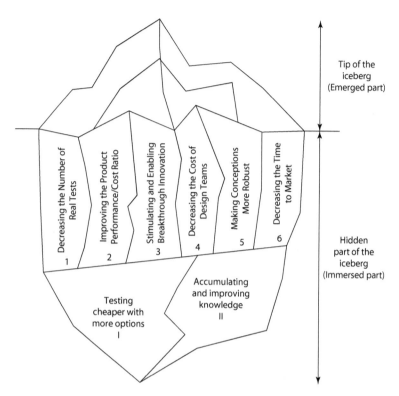

FIGURE 3.1
Modeling iceberg: emerged and immersed parts.

2. *Financial benefits*: These are the indirect benefits, the consequences of
the intrinsic ones, that in the end offer financial returns (via cost cuts
or increased income).

We're going to clarify these two types of benefits in the next section
(Figure 3.2).

3.2.2.1 Two Intrinsic Benefits

3.2.2.1.1 Intrinsic Benefit 1: Testing Cheaper with More Possibilities

As explained, simulation enables one to perform virtual tests with major
advantages compared with the live experiments we'll detail below. The limits of "virtual" tests will be presented in Section 3.3.

- *Testing cheaper*: Live experiments are often expensive as they require
  a real version of the system to be tested (products or even building
  prototypes are needed) and means to handle these tests (both measurement equipment, sometimes test bench, but also engineers and

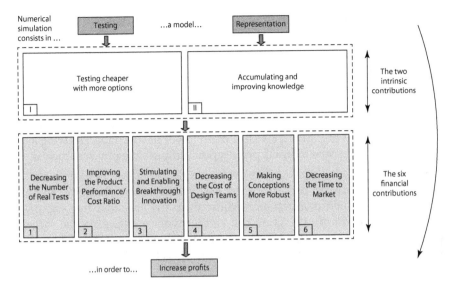

FIGURE 3.2
Levers of value creation in numerical simulation.

technicians who will proceed to the tests). "Virtual" tests, by nature, do not need these resources and are consequently less expensive (we'll detail their cost in Section 3.3).

To manage a single test, then, simulation is often much less expensive.

- *Testing with more measurements*: It is often difficult to access some measurements with physical tests for two reasons: technical achievability and cost. To illustrate the first point, during a test performed on a vehicle for example, it is not possible to calculate the stress inflicted on the wheels because the measurement would be too disruptive and would disturb the test. For achievability reasons, only a limited number of measurements are possible. On the other hand, in simulation, almost all measurements are possible, as the test is virtual (the limit being the detail level of the model and the capacities of the numerical tool). This can also be explained by cost. In a test, every additional measurement has a cost (if it requires extra equipment), whereas in simulation, this extra cost is insignificant.

  Simulation allows one to access a wider number of measurements than physical tests do.

- *Testing the untestable in reality*: In simulation, all that is required is the system must be conceived and defined to be able to be tested, whereas in real tests, additional constrictions appear. Let's mention two common constrictions. The first and obvious one is that to be physically tested, the system has to exist. The second one is danger, many tests being nonperformable because of it.

So, simulation allows one to perform tests that are not achievable in reality.

- *Testing faster*: In simulation, tests can be performed faster than in real time. For example, it is possible to simulate 10 years of the natural use of a screw within a few seconds of simulation. The time requested to organize the test is often lower in simulation as well, as it is not necessary to build prototypes and possible physical means.

- *Testing and being able to test again*: In simulation, it is very easy to perform a test twice, or perform an equivalent test by changing some settings, which is much more difficult in the case of a live experiment. Performing the same test again involves assuming the same costs. And changing a simple parameter, like the loading of a truck or the outside air temperature, is much less trivial for a live experiment than for a virtual one.

We'll finally add two more advantages that are consequences of the ones mentioned above.

- *Testing earlier*: We've seen that simulation allows one to test a system if it doesn't yet exist. During the conception process of a system, it is possible to test it earlier through simulation. For instance, there is no need to wait for a rocket or a prototype to be built to test it virtually.

- *Performing more tests*: All the previous points have as a natural consequence that when simulation is expanded, the final amount of tests (both virtual and live) increases dramatically. This has positive impacts to be covered later.

In a nutshell, simulation not only allows cheaper experiments than live ones but also offers a large number of extra possibilities. This first underlying benefit of simulation has multiple financial consequences, detailed after the second intrinsic benefit.

3.2.2.1.2 Intrinsic Benefit 2: Accumulate and Improve Knowledge

We have seen earlier that a model is a representation of a system. This has two consequences:

1. *A better understanding of the system*: To represent and model a system, it has to be understood. This means identifying its inner mechanisms in order to predict its behavior. Consequently, this activity needs to accurately grasp the behavior of the system studied. When simulation is not used and tests or expertise are appealed to, this fine comprehension is less necessary and less developed. It should

be mentioned that this improvement of knowledge is a side effect, coming out of necessity, and that it is not the initial purpose. Its impact will be covered.

2. *Accumulation and sharing of expertise*: A model is a very good tool to accumulate expertise:

   i. First, the model necessarily includes the expertise of the system. It is the translation into numerical form of the known physics that rule its behavior. With time, the comprehension of systems improves, and models are often updated for a better prediction of their behavior. This update is again done by necessity, otherwise simulation becomes obsolete (in opposition to written documentation that requires constant effort, is unnecessary in the short term, and isn't often made on a regular basis).

      Finally, a model is a very short way to store expertise. The numerical model only gathers equations that are imperative to predict its behavior. This language is generally denser and more efficient than a written document. It goes straight to the point.

   ii. A model is also a great tool to convey knowledge that has been capitalized on. On the one hand, it goes straight to the point and its language is universal. (Whether they are mathematical or numerical, the languages of modeling are much more universal than the languages of different countries). On the other hand, it is testable; that is to say, it is possible to run simulations by changing the hypothesis and observing the obtained results (internal variables being observable in opposition to a test). Learning through tests/mistakes can be done and is often efficient.

   iii. Therefore, whether it is to capitalize on or to convey expertise, a model is often much more efficient than traditional methods. Let's take the example of an industrial company designing nuclear plants. Models of thermohydraulic water circuits are finally one of the solutions to gather expertise. If new discoveries occur during turbulent floods, they will *in fine* be integrated into the model. After 20 years of R&D regarding water circuit floods, the models are often the most updated and condensed, to capitalize on and convey expertise in the company.

This can be summed up this way: "A model is the crystallization of the know-how of a team" (Manuel Tancrez).

The combination of both effects leads to the improvement of the expertise within the company. The comprehension of the systems is finer and is better capitalized on and shared. The other core benefit we detailed previously (cheaper tests with more possibilities, via an increased number of performed tests) also comes to enrich the level of expertise.

3.2.2.2 Six Economic Contributions

Let's see how these two intrinsic contributions of numerical simulation are connected with six economic contributions that eventually lead to profit for the company. Every economic contribution to be covered here is directly or indirectly a consequence of the intrinsic contributions.

For each of these six contributions, we'll add an indicator of

- *Value creation potential* ($/$$/$$$)
- *Short-, middle-, or long-term aspect* (ST/MT/LT)

These indicators are nothing but a tool allowing one to get an idea of probability, and are obviously not to be given the upmost consideration.

According to the company and the field of industrial activity, simulation may be used more or less, and the six contributions detailed below may be more or less significant. We will use examples, but to respect confidentiality, the name and associated operating numbers will not be mentioned.

3.2.2.2.1 *Economic Contribution 1: Decreasing the Number*
       of Real Tests ($$$, ST/MD/LT)

Testing cheaper with more opportunities opens a pathway to supplanting live experiments with simulation and *in fine* decreasing their related costs. This argument is often the first mentioned (as it is easy to understand and assess) when backing investment in numerical simulation, but we'll see that this needs to be tempered.

Three types of live experiments can be substituted by simulation.

1. *Tests for the Design*: These are run to support a product design. For instance, to design an aircraft engine, prototypes are built and tested in order to get rid of uncertainties: Does the engine reach or exceed the temperature limits? What will the engine performance be with the planned dimensions? and so on. These tests are often very expensive: they require building a prototype on the one hand and the mobilization of operating means on the other hand (industrial equipment, technicians, and engineers), not to mention the common hazards. To sum up, tests of conception campaigns, including the production of a prototype, often cost millions of dollars or even more.

   These tests may sometimes be avoided, at least partly, in favor of numerical tests—that is to say, simulation. Here, the aircraft engine is designed considering hypotheses in compliance with the progress of the art of conception, and simulations are performed with the objective of obtaining the same results as with live experiments (this alternative through simulation also has weaknesses, detailed in Section 3.3).

2. *Tests for Settings and Calibrations*: These are for final adjustments. The majority of complex industrial products require a calibration

phase. As predicting the behavior of these products is a complex matter, degrees of latitude are often admitted, and the products are arranged at the end of the conception process.

For example, in the case of a vehicle, the amount of fuel injected depending on how far the gas pedal travels is adjusted once its conception is achieved.

This is only one example among countless parameters adjusted. The tests also induce serious costs. The prototypes are generally less costly than those built in the design process (as they are more similar to the final product), but the number of tests to be made is often higher.

These tests can be avoided by substituting it with virtual adjustments based on simulations. As in conception, the product is digitally designed, and only then are the adjustments performed through virtual means rather than real ones. These models are often more precise than in conception, as more experience is collected during the preceding step. Also, even if setting a specific adjustment may not be possible through simulation, it is often convenient to manage the transition by setting up a digital premeasurement process. So, the adjustments are presettled with simulation, and the purpose of the tests is to check the settings or adjust them, which allows one to dramatically reduce their number and therefore the costs.

3. *Tests for Validation*: These are performed to make sure the product has the expected features. Going back to the aircraft engine example, the purpose will be to check that the engine performance, in its final version, meets the requirements of a market launch. Once again, these tests are very expensive. Then, the products tested are those to be sold (and so they are less costly than prototypes), but the number of tests may be very substantial, and the measurement conditions must be very precise and comprehensive.

The tests may be warded off by validating the products via virtual tests. These tests are considered the most challenging to replace, which is often true. First and foremost, as this subject may be touchy, the impossibility of testing every feature must be kept in mind. A selection will have already been made. Some features will not have been tested, and industries rely on expertise to guarantee that they will be operational. The purpose of simulation is to extend the range of features that will not be tested in reality; there is nothing new about that. Then, in order to decide what validation test to use instead, it is necessary to figure out the expected accuracy. As will be covered in Section 3.3, modeling is very helpful to predict *relative* but not *absolute* behavior. Thus, the most easily replaceable tests are those testing relative behavior.

We pointed out these three types of tests for their potential benefit, and the ease of implementing simulation may vary. So it is common to initiate a simulation by targeting only one of these tests (the most profitable, which depends on the context) and then extend it to the other two.

The potential reward combined with the decrease of the tests number is quite substantial. They represent up to several dozens of percent of the total amount of the expenses of many companies. Most industry leaders involved in this work have already undertaken cost cuts thanks to simulation, but there is still a long way to go.

Also, these profits may be made in the very short term, via the immediate removal of certain tests, and are generally kept in the long term (because for each generation of products, the test reduction is maintained).

Nevertheless, these assessments must be tempered as these cost-cutting potentials are usually overestimated. Why?

- First, a total removal of the experimental phase in order to supplant it with a simulation is common. The exact accuracy of the simulation is very often misestimated or wrongly forecast, then either additional tests are scheduled in an emergency or side impacts are reported (e.g., quality impacts, additional costs, delays, delivery issues).

- Second, simulation sometimes needs additional specific tests (covered in Section 3.3.1). These result in costs that are sometimes not forecast.

However, these two threats usually occur during the changeover time (when a test phase is completed). The projects to come that will profit from this test phase saving will, most likely, not be impacted by the threat we have just mentioned.

Our main message here is that financial profits are often actual but over a longer term than expected. In this way, despite the true potential for cost cuts thanks to simulation, these are often overestimated, at least in the short term.

3.2.2.2.2 Economic Contribution 2: Improving the Product Performance/Cost Ratio ($$$, MT/LT)

In any industrial company, choices need to be made during the product conception process. They come from a compromise between costs (cost price and conception cost), product performance, quality, and deadlines. These choices may be more or less well optimized.

Simulation often enables one to dramatically optimize the design bias through two effects we are going to cover.

1. *Choosing among a larger panel*: Simulation broadens the field of possibilities and in this way betters the final decision suitability. Why is the realm of possibility widened? First, we saw that simulation allows one to increase the number of tests for an equivalent cost. Thus, for an equal expenditure, more solutions may be considered. We also mentioned that simulation enables one to assess solutions impossible to get through live experiments. Finally, simulation

permits one to design earlier (the reasons will be explained in Contribution 4) and to choose among possible solutions that won't be reachable later. Indeed, any company designing a product has to set conception decisions over time. The more time goes on, the more the range of possibilities reduces. The opportunity to test and design sooner enables one to minimize this field of opportunity reduction.

2. *A better choice*: The inner contributions of simulation boost the understanding and knowledge of systems, allowing, with an equal solution sample size, a better choice.

Therefore, simulation supports decision-making improvements within the company. How can it be profitable? The impacts are either connected to a product cost (price cut) or to an income increase (performance and quality development, doping sales in the middle and long terms).

Besides, the encouragement of modularity allowed by simulation has positive repercussions on the product performance/cost price ratio (how simulation contributes to modularity will be covered in Contribution 4). Indeed, modularity curbs cost prices (as the number of components is reduced, their unit price lowers—in particular, through a toughened power balance with suppliers) or raises the performance and the quality of products (with a smaller amount of components, their control and their features widen, with, in the end, a positive impact on the products).

Many examples could be mentioned. In the car industry, when a component selection process is coming up, it is more and more common to call on simulation to select a part or a supplier. Revising cost prices in an industry where margins are very tight is very profitable.

The financial stakes of these effects may be decisive. However, the earnings are seldom made in the short term. They can be harvested in the middle term thanks to cost price reductions, and mostly in the long term through product performance and quality improvements.

3.2.2.2.3 *Economic Contribution 3: Stimulating and Enabling Breakthrough Innovation ($$$, MT/LT)*

Using simulation favors innovation incentives within the company. Two types of innovation can be identified: incremental and disruptive. We have already stated how simulation enables optimization (it is conceptually close to incremental optimization), and in the end, the financial return on existing products. We will deal with disruptive innovation in this section.

Disruptive innovation is stimulated through two mechanisms that both profit the underlying assets of simulation.

1. *Promoting innovation*: Simulation engineers usually comprehend systems in a more theoretical way, and this new point of view, free from practical restraints, generates new ideas. Solutions that were never considered often come up. This approach, which is more open and

is the outcome of better understanding, generates a culture full of innovation.

2. *Making innovation achievable*: We saw that simulation permits one to perform tests more cheaply and with more opportunities. This makes the execution of innovation more practical through three determinants we're going to detail. First, simulation enables one to elude many technical achievability problems. For example, product conception is sometimes prevented because of the inability to make live prototypes.

   Hence, innovation development costs may also be prohibitive without simulation. A large number of tests are carried out on prototypes to validate innovative concepts, and their costs may become proscriptive if actual tests are performed. Finally, the execution period may be a pitfall for an intended innovation. In chemistry, many projects are aborted because the development time is considered proscriptive, whereas simulation sometimes favors condensing them (we'll cover the reasons later).

To cite only one example: The development of driving assistance for vehicles, which became widespread in the early part of this century. Simulation clearly stimulated developments that wouldn't have existed without it. Among other things, these numerical tools authorized a wide range of situational tests that vehicles may come across during their service lives, including extremely dangerous ones to perform in reality.

In conclusion, this innovation advance has a monetary impact as new products are permitted to be developed, potentially resulting in significant revenues for the company. However, this payback can only be felt in the middle and long terms.

3.2.2.2.4 Economic Contribution 4: Decreasing the Cost of Design Teams ($, MT/LT)

Simulation permits one to reduce the cost of design teams according to two main points:

1. *Reducing the number of conception studies (for an equal result)*: This is the repercussion of three subfactors:

   a. *Conceiving earlier*: We saw that the basic advantage of simulation was the ability to perform earlier tests based on virtual methods. As a result, the conception of the system can be anticipated. Decisions are able to be made earlier, the number of manipulated hypotheses reduced, and thus the number of studies channeled. This is how the step forward in the design process reduces the number of studies. Contribution 2 also outlined that this has a profitable impact on the product performance/cost price ratio. These contribution are illustrated in Figure 3.3.

b. *Stimulate modularity*: As mentioned before, simulation allows one to choose among a larger panel of opportunities and make a better educated choice, boosting modularity growth. With a larger panel selection, it is easier to consider a modular component (a component already selected for another project). Also, with better expertise, it becomes easier to determine whether using that modular component makes sense in this specific context. Simulation then becomes a strong ally to implementing modularity within the company. The virtuous effects of modularity are dual. First, it allows one to reduce the number of studies performed by the teams, as the components in question are already known. Second, modularity contributes to improving the product performance/ cost price ratio, an aspect already covered in Contribution 2.

c. *Making conceptions more robust*: We'll cover this point in more detail in Contribution 5. Only one of the consequences of robustness has to be underlined: avoiding extra studies when all decisions taken in the past are overhauled.

2. *Reducing the average cost thanks to a conception study*: There are three reasons for this. Firstly, better expertise, as already explained, improves team efficiency and, consequently, the average cost of the studies. Secondly, simulation radically reduces document resources. Indeed, as detailed in Contribution 2, models efficiently compact a large amount of information about the system studied, and less information needs to be gathered into long written documents. In certain cases, when it comes to specifications, models can also

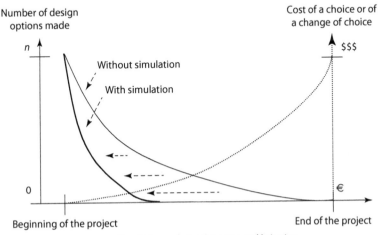

Simulation reduces the costs and bringing
the design decisions forward

FIGURE 3.3
Reducing the costs by anticipating the conception.

remove documentation phases too. Lastly, it is easy to reuse a model for a similar study. Changing a parameter is accessible, whereas it is not for a live experiment. These three reasons reduce the average cost of conception studies.

As a result, fewer and cheaper studies generally result in cheaper conception costs. The profits, even when they are high, are usually lower than the benefits from the contributions detailed earlier. These profits are made in the middle and long terms. Many companies have them, even if they are difficult to measure.

3.2.2.2.5 *Economic Contribution 5: Making Conceptions More Robust ($$, MT/LT)*

Numerical simulation is a precious tool to make conceptions more robust. As a matter of fact, we have seen that it first allows one to increase the number of tests, resulting in progress in detecting potential problems, and to check the robustness of the operating choices over a wider range of stresses. Secondly, a better system understanding and appreciation improves the efficiency of identifying risky situations that may occur in response to a specific application. Simulation begins to then identify the predictable critical scenarios in a more reliable and exhaustive way.

This upgrade of robustness has three consequences:

1. *Cutting costs of the conception teams*: We have already covered this point in Contribution 4.
2. *Decreasing the costs of nonquality*: A stronger conception provides a better product quality that prevents nonquality cost issues—for instance, the recall campaigns of after-sale service, which happen to be very costly.
3. *Allowing certifications*: Some standards and certifications require a level of robustness that is unlikely to be achievable without simulation. Let's mention the ISO 26262 standard, the application of which is very unrealistic without simulation. The company is impacted in a way that the revenues coming from the certification would be missing.

This gain in robustness brought about by simulation can be related to major economic stakes, although it impacts the company only in the middle or long terms.

3.2.2.2.6 *Economic Contribution 6: Decreasing the Time to Market ($$, MT/LT)*

Numerical simulation is an efficient tool for reducing the *time to market*, a consequence of the two following points:

1. *Designing earlier*: This point has been covered by Contribution 4.

2. *Designing faster*: This has also been covered by Contribution 4. Simulation cuts down the number of studies (for an equal result) and lowers the average amount of necessary work for each study. The design time consequently decreases for an equivalent outcome. We shall also add a last feature that can make a difference. According to Intrinsic Benefit 1, simulation enables one to test faster than by performing live experiments and so reducing the design process time.

Thus, by enabling one to design sooner, the time to market is shortened, with major monetary impacts. Firstly, it may become a competitive asset, because the products would, for example, better fit the current market expectations than competitors' products do. In addition, this may generate quicker incomes.

These profits will be saved in the middle or long terms.

3.3 Simulation Costs and Limits

After having covered the benefits of simulation, let's focus on its costs and limits.

3.3.1 Costs of Simulation

Simulation generates four sorts of common costs (which are commonly either overestimated or underrated):

1. *Simulation engineers' salaries*: This is generally the prominent cost item. It is nothing but the salaries of the engineers who build the models and draw the results. Throughout the simulation process, engineers often seize time from designers' schedules (this may be considered in-house training on the studied system). To be completely exhaustive, a small amount of nonsimulation engineers' salaries could possibly be cut (but this would be a temporary measure).

2. *Live experiments*: This point may look surprising. Actually, simulation requires one to perform additional specific tests. These simulation tests need to be frequently performed on subsystems to fully understand their functioning or on the global system to carry out correlation work on the model. Of course, it is often about performing, for example, five extra tests, to save doing fifty later on.

3. *Software costs*: Numerical simulation involves using the software adapted to the field. Most software on the market is paid for (and frequently includes after-sale and technical support services). These

costs can even reach the salary charge of the engineers using it, knowing that several types of software are sometimes mandatory.

4. *IT tools*: This is about computers that help with calculations (those computers are often more expensive than standard computers). These computers can become specifically dedicated servers, becoming even more costly.

3.3.2 Limits to Simulation

We have spent a long time detailing the benefits of simulation. To be as impartial as possible, and to lead to a rational decision about investing in simulation or not, let's deal with its limits.

We mention here only the two main limits to numerical simulation compared with traditional methods and not the problems linked to its expansion (to be covered in Chapter 4), where the matter is to determine whether it is profitable or not (to be studied in Section 3.4).

These limits may reduce the application scope of numerical simulation and sometimes increase the costs (also to be studied in Section 3.4).

3.3.2.1 Limit 1: Precision Level of Simulations

A simulation result is always an approximation. Indeed, we previously mentioned that a model was only a representation of reality, a simplification. So, a simulation result will only be an approximation of reality, but this obviousness is often overshadowed. This is not prohibitive.

- *Necessary level of confidence to make a decision*: In companies, no decision is made on the basis of absolutely accurate studies. We have to operate with information that isn't 100% reliable.

- *Trust level of live experiments versus simulation*: Live experiments, often the alternative to simulation, are also approximately accurate. This second point is often forgotten. Anyone who has ever performed a test knows that measurements are imprecise, that the means are not perfectly representative, and that the conditions of the tests can't be absolutely controlled. So, a live experiment result is always an approximation.

  We can say that the accuracy of live experiments is often overestimated, whereas it is underestimated for simulation: "Those who trust the results of a numerical model are those who did it. Those who trust the results of a live experiment are those who didn't" (Benoît Parmentier).

Lastly, is the inherent inaccuracy of simulation a limit?

- Yes, if the accuracy required for the study in not technically achievable (in this case, please refer to Limit 2)
- Yes, if the accuracy is not controlled (Please refer to Chapter 4, which explains how to avoid this situation.)
- No, otherwise

This first limit confines the field of applications where simulation adds value (see the following section) and requires best practices for its use.

3.3.2.2 Limit 2: Technical Feasibility of Simulations

Simulation may not be technically possible. These barriers can be of two types:

1. *Mathematical and numerical*: Some complex mathematical equations haven't yet been efficiently solved. In this case, the progress made in mathematics will show the solution. Also, the limits are sometimes numerical and they are bound to the power of the computer. Let's mention a large number of models of combustion that require simplification in order to be used with reasonable calculation delays.

2. *Knowledge level*: The current knowledge of physics in the studied models is a constraint on simulation. To be modeled, a system has to be represented. In some cases, the phenomena are wrongly understood and it is impossible to create the model. The system's inner mechanisms can't be represented.

   This second limit, again, restrains the simulation field of application, or at least its efficiency. (The impact on costs and/or deadlines is to be covered in the next chapter.)

3.4 Deciding Whether to Use Simulation or Not

These two situations are frequent in companies.

1. Not using simulation in a specific field and missing high value potential
2. Using simulation in another field for a very limited profit, turning it into a failure

It is then about discerning the situations in which using simulation is smart: How?

We previously established that using numerical simulation in a company is an investment that

- Generates profits (Section 3.2)
- Has a cost (Section 3.3)

To decide on numerical simulation, the main question is whether investment is profitable or not. (It can also be pertinent to know the level of profitability, its short- or long-term aspects, and the level of risk, all the while keeping in mind the expectations.)

The main obstacle to answering this question consists in providing a budgeted answer (in terms of profits and costs). This obstacle is often overestimated. The point is not to get an exact result; it is about assessing the order of magnitude of the potential value creation in the most impartial way.

In this chapter, we have given leads to estimating these orders of magnitude and comments on how to escape disappointment. It may be relevant to partner a simulation expert with a top management representative in order to briefly evaluate the potential outcome of a simulation. Simulation engineers often quite naturally overestimate the outcome of their involvement (or they believe in their value through a specific input when another one they didn't think about is much more consequent). Management conversely often underestimates the outcomes of simulation. So, it can be dangerous to believe in misinformed intuition when making the decision on such an investment. This is why this chapter has been split, to make decision-making as impartial as possible and to boost simulation profits. Our goal here is to make that decision as rational as possible (even if a right decision for wrong reasons happens or vice versa).

A pertinent use of simulation can easily generate profits 10 times higher than the costs. Investing progressively in simulation to reduce risks is also commonplace. This way, expenses are low and the first success of the initiative encourages the development of the simulation extension.

At this point in the book, the reader holds all the cards to decide on simulation investment. If the answer is yes, now is the time to wonder how to expand numerical simulation. Chapters 4 and 5 are devoted to answering that question and to detailing the key success factors to obtain the right results.

4

Efficient Use of Numerical Simulation: Technical Aspects

What are the technical best practices for using numerical simulation efficiently? There are many types of numerical simulation models but with several similarities; a certain number of best practices are universal.

4.1 Different Kinds of Numerical Simulations

There are various kinds of models, and their applications are diverse. We'll define here the main kinds in order to put things into perspective before dealing with the technical best practices.

- *Dynamic or static*: A model is dynamic when its state variables change over time (these models are mostly described by differential equations). The ice cube model in Chapter 1 was a dynamic model. A model is conversely static when it does not depend on time. For instance, a model assessing the intensity of an electric current passing through an electric resistance, a function of the tension delta at its terminals, can be static (according to Ohm's law, the current at its terminals is proportional to the voltage delta).

- *Discrete or continuous*: A model is discrete when its internal variables only change a finite number of times. For example, modeling the number of people inside a building can be done in a discrete way (the number changes when someone enters or exits the building). On the contrary, a model is continuous when its state variables change in a continuous way (in time and value). For instance, the melting of the ice cube model in Chapter 1 was continuous, as the thickness of the cube was changing continuously.

- *Determinist or stochastic*: A model is determinist when the described behaviors are fully predictable and do not rely on random phenomena. Rebooting the same simulation will provide an identical result. The ice cube melting model was determinist because the included equations didn't involve random effects. Conversely, a model is stochastic when it includes behaviors considered probabilistic. A very simple example for that would be a dice-rolling model, which would then include a probabilistic distribution.

- *0D or 3D*: A 0D model (also called a *system model*) describes physical phenomena without any consideration for the dimensional variables' influence on the state variables. For example, the previous ice cube–melting model was one of them: it didn't include dependence on axes x, y, or z. A 3D model (also called a *dimensional model*) has the tridimensional dependences intervening. For example, if we had complexified the ice cube model, we could have made a 3D model to highlight the nonhomogeneity of the ice cube melting, and so its reliability to the axes x, y, and z.

  Also, the notion of a 1D model, which only uses one geometrical dimension, is often associated with the 0D simulation (because of its abstraction level). The 2D simulation is, on the contrary, often linked to the 3D simulation.

- *Command or organic control*: A control command model depicts the logic of the control that is or will be implemented into an actual calculator. For instance, a calculator regulating the fuel injection in an aircraft engine can be modeled. Conversely, an organic model represents the physics of the mechanisms impacted by the controller. For example, a reactor's combustion chamber can be modeled to assess the power generated by the combustion of the fuel injected. Control models are often paired with organic models.

- *Mono- or multiphysical*: A monophysical model includes phenomena related to a specific field of physics (thermics, mechanics, etc.). For example, the ice cube–melting model was one of them, but only the thermic equations were used. Conversely, a model is multiphysical if several laws from physics are involved. In this case, the phenomena at stake and their time constants can be very different. For example, the modeling of the electric power of a wind turbine can be made in a multiphysical way (solids and fluid mechanics to assess dynamics, electricity to assess the power generated).

- *Process, functioning, or hold*: A process model's purpose is to predict the system's production performance. For example, a chemical model can be designed to optimize the manufacturing process of a chemical product. A functioning model's purpose is to forecast the system's efficiency when it is used. For example, an aircraft engine model allows one to predict the thrust it will have to deploy. A hold model's purpose is to forecast what stresses will cause system failures. For instance, a mechanical model of a car chassis will allow one to determine what stresses can break the system or change its properties.

Therefore, the classifications of the models are very varied. Of course, some specific techniques have been designed for each kind of model, and if needed, the reader may refer to specific technical publications. Several best practices are common to any numerical model and are often misunderstood. These common best practices will be covered in this chapter after defining the five main steps of numerical simulation expansion.

4.2 Five Steps to Expand Numerical Simulation

We previously explained there are many types of models. But like generic best practices, the main steps of the building and use of models can be defined generically. We'll detail these here, using the example from Chapter 1.

1. *Defining the aim of the simulation study:* This is about precisely defining the expected aim of the study. So, with the example from Chapter 1, the ambition is to assess the ice cube melting. The data to be specified in the context of that study are the cube size, the temperature of the water in the glass, and so on. It is necessary to tally all of that with the problem.

2. *Building the mathematical/physical problem:* This is about performing the modeling work, or building a conceptual model of the system, then translating it into equations. With the example from Chapter 1, the result of this step is

$$\begin{cases} \dfrac{\partial c(t)}{\partial t} = -\left[A + B \dfrac{1}{c^2(t)} \right] \\[2ex] A = \dfrac{5h_{water}\left(T_{water} - T_0\right)}{3L_{water\ fusion}\, \rho_{ice}} \\[2ex] B = \dfrac{\varphi_{solid}}{3L_{water\ fusion}\, \rho_{ice}} \\[2ex] c(t_{final}) = 0 \end{cases}$$

3. *Converting the model into a numerical model:* This is about translating the equations into numerical calculations that can be used by a computer. This step may be more or less open to the user, depending on the software used. With the example from Chapter 1, it gives the following:

```
T_water = 273.15 + 20
T_0 = 273.15 + 0
h_water = 250
L_water_melting = 334 *10^3
Rho_ice = 917
Phi_solid = 1
A = (5 * h_water (T_water - T_0 ) ) / (3 * L_water_
melting * Rho_ice )
B = Phi_solid / (3 * L_water_melting * Rho_water )
c = 0.02
t = 0
Delta_t = 0.1
While (c > 0)
    c = c - Delta_t * (A + B / (c^2) )
    t = t + Delta_t
End
Print t
```

4. *Producing and delivering results*: This is about approving then using the numerical model to find and deliver a final answer to the problem defined in step 1. This way, in the Chapter 1 example, the answer is "8 min."

5. *Storing the model and its results*: This is about storing the model to potentially use it again later, as well as the results in order to guarantee traceability.

We'll see in the following chapters that, depending on the model used, many common mistakes happen. These are both technical and organizational mistakes. Being aware of the best practices enables efficient and relevant work; this will be explained in the following sections.

As mentioned in the Chart 4.1, the technical best practices we are about to detail are mostly related to the modeling step, whereas the organizational best practices to be covered in Chapter 5 are transverse in regard to these steps.

4.3 Eight Technical Best Practices

Eight technical best practices can be listed. To show them in a clear and understandable way, we have decided to introduce them with a case study. Thus, through Chapters 4 and 5, we'll follow the experiences of Doug, a junior simulation engineer on a specific case. He's going to set a modeling procedure on a new topic and will commit eight common mistakes; he will

		Technical best practices	Organizational best practices
Simulation steps	A/Defining the aim of the simulation study	1	
	B/Building the mathematical/physical problem	2 and 3	
	C/Converting the model into a numerical model	4 and 5	1 through 8
	D/Producing and delivering results	6 and 7	
	E/Storing the model and its results	8	

CHART 4.1
Connection between the steps of simulation and the technical and organizational best practices.

then solve them by applying the eight associated best practices. In this way, each best practice will be preceded by a short background in order to reveal the common mistake.

The case reported here will be more complex than the one used in Chapter 1 (ice cube melting) and more representative of a true issue. However, we will not detail the entire study, to avoid making the book too cumbersome. We will only show the key points, exposing the frequent mistakes and the best practices. Finally, Doug's mistakes may appear clumsy: they are amplified in order to finely detail each best practice. But the reader should not be mistaken: from the moment the situation becomes more complex, it becomes difficult to take a step back, and even experienced simulation engineer can make these mistakes!

CASE STUDY BACKGROUND

Doug just got hired by a leading company in fridge manufacturing.

Project manager: "We are working on a new type of fridge. We intend to develop a varying-speed compressor for our fridges, in order to substitute the current constant-speed compressors. A first prototype will be available within a year, but we must make sure right now that this compressor change will enhance the performance, in order to decide whether to keep on with the project or not. I would like you to build a numerical model of the fridge within 2 months."

Doug: "Great, I'll get to work right now."

4.3.1 Best Practice 1: Defining the Objective

MISTAKE 1

Doug just got hired, he wanted to do things right, and started right away on the modeling he was assigned. Time flew on, and a week before the deadline, he met the project manager and let him know about the progress.

Doug: "I have built the model. I'm able to assess the fridge's performance, meaning the necessary duration to cool down a food item and how much electricity it uses."

Project manager: "Very good, but only the energy use matters; the cool down duration is not at stake here."

If he had known, Doug could have saved work and built a simpler model.

Project manager: "So, by the end of the week, thanks to your model, you need to give us an assessment of the fridge's energy use in a standard real-life situation: a 24-hour day, with two door openings, and eight food items inside."

Doug: "I didn't know we had to study situations with door openings! This involves circumstances I didn't take into account. I need two extra weeks to do this job."

Project manager: "These results will be outdated in two weeks. We need to make the decision by the end of the week. So we will base our decision on expert opinion, without simulation."

Thus, by not clarifying the objectives, and even if Doug had built a model that could have been more relevant for another matter, this one wasn't appropriate for the current problem and didn't provide any value in the end.

To avoid this mistake, the aim of the task has to be defined. The best practice we are about to review is very simple but very scarcely applied. The key points to properly define the aim of a simulation are the following:

- *Understanding why the study needs to be done*: This is the starting point. It looks obvious, but the purpose of the study is often forgotten when the situation becomes more complex. Understanding prevents one from doing useless work (a few *why* questions sometimes help to realize a study has no relevant reason to be done.)

- *Defining the result(s) variable(s)*: The expected data outcomes. Doug's example deals with energy use. The result variables must be separated from other internal variables that may also be requested. Indeed, many variables may be requested, mostly to understand some specific phenomena or check some behaviors. In this case, the expectations regarding these internal variables will be low (especially in terms of accuracy; see the following point).

- *Defining the expected accuracy of the variable(s) to observe*: This point may seem difficult to define, though it is crucial. In our example, Doug will have to precisely define the expected energy use (in watts) with the project manager. The expected accuracy of the results is sometimes difficult for management to define (project managers tend to request perfect accuracy). As we mentioned, a model is always an approximation, so the result is imperfect. Even before initiating the study, the simulation engineer must define whether the imprecision of the model fits the project manager's expectations. There are two possibilities: either the need for precision is conflicting, so a pointless simulation study must not be initiated, or the need for precision is adequate, and the project manager will determine the necessary precision level of the model. Even if this step requires several hours of meetings and negotiations with the project's management, it prevents potential backward steps, which are often very costly.

- *Defining the system(s) to study*: What system is studied (in our example, the fridge) and its features (e.g., size, power, etc.). The system can be studied through several possible filters.

- *Defining the life situation(s) to study*:
  - *Defining the test environment*: All that surrounds the system (the fridge) that may interfere with the study results. For instance: What is the outside temperature? Is the fridge heated by an outside radiation (the sun)?
  - *Defining the stress(es) of the system*: What are the parameters of the expected test(s) and what is its initial state? For example, is it about assessing the average 1 h energy use with the door closed, or is it about a year's average energy use with multiple door openings? Regarding the initial state: What is the temperature inside the fridge? Does it contain food items?

- *Defining the deadlines*: This is a typical project management issue, but it is good to be reminded. Indeed, deadlines are often hazy at the beginning of a simulation process (or at least the expectations are; see the previous points). Yet, agreeing with the deadlines is vital to ensure that, once again, the works will be useful and relevant when delivered.

- *Defining the costs*: This is also a classic in project management. It is important to agree with the orders of magnitude before initiating the studies. For example, if the simulation study requires some expensive tests, its achievability must be checked at the very beginning.

So, making the objectives of a study very clear is vital, due to the preceding points, even before initiating work. This may seem tedious or even useless because it looks obvious. However, experience reveals that many mistakes are prevented by using this best practice, and it is extremely profitable to do so.

A few observations before stepping into Best Practice 2:

- *Similarity with live experiments*: The reader may observe that the points covered so far may apply to live experiment bill specifications. Indeed, as described in Chapter 1, for a project manager, simulation is just an alternative to live experiments. The specifications should be alike. However, it is clear that the expectations are generally more thoroughly described for live experiments than for simulations (we'll see why in Chapter 5). Thus, the best practices for the test definitions must also apply to simulations.

- *Cancellation of the study by defining the need for simulation*: Frequently, for reasons mentioned previously, making the need for simulation clearer leads one to cancel it. This mustn't be understood negatively. When it happens, it means that otherwise the study would have been initiated and then canceled without results.

- *Starting work when the objective isn't crystal clear*: The previous point mustn't lead, if clarifying the need is not easy, to an anticipated cancellation (this is absolutely not our message). Some of the previous points can't be defined at the beginning of a study. For example, it may not be possible to decide on a model's precision before starting work. In this case, our message is not to not initiate the study. It is to choose whether to initiate it or not, assuming that it may not succeed. Depending on the situation, the decision will vary.

- *Iterations*: To continue with the previous point, it is frequent that iterations of the problem definition will occur. A first objective is set at the beginning of a project, and then, depending on the breakthroughs, some others can be added to the initial one. The problem definition is enriched with the work's progress.

- *Generic model*: In most cases, when a model is built, it doesn't meet only one need. Many studies will be done with a single model, with several variables requested afterward. Does the simulation need the definition step to remain mandatory in this case? Right from the beginning of the project, it is about identifying the whole model's needs in order to make sure they will really be taken into account in the model construction process.

- *Already-existing model*: Frequently, when a simulation is requested, a model already exists and may be reused. The previous warnings remain applicable, especially as updates or complementary works may occur.
- *What is requested versus what is possible*: Commonly, the final definition of the study objectives come from a balance between the initial need and what is achievable. On the one hand, some parts of the demand appear to be irrelevant, some are extremely difficult or costly to do, and some are just not mentioned, as they are considered to be too complicated to express. On the other hand, means are not unlimited, or points that were not expressed may come up especially when they are easy to address. Thus, the definition of the study becomes more precise after the first discussions and may deepen along the way.

APPLICATION OF BEST PRACTICE 1

After talking to his project manager, Doug defined the objective of his simulation.

- *Understanding why the study needed to be done*: The new generation of fridges had to be more efficient in terms of electricity use. So the purpose of the study was to predict the power use of the future fridges compared with the old ones. The first prototype would be available a year on, but the viability of the project had to be determined soon because it was costly. That was why the simulation request had to be done immediately.
- *Defining the result(s) variable(s)*: The energy use.
- *Defining the expected accuracy of the variables to observe*: ±20%.
- *Defining the system to study*: All the features of the fridge (regarding the data to be useful later: electric compressor features, etc.).
- *Defining the life situation to study*:
  - *Defining the test environment*: 20°C outside temperature, no sunlight.
  - *Defining the system stresses*: The system was studied over a 24 h day, with two door openings and eight standard food items inside.

- *Defining the deadlines*: The final results had to be available within 3 months.
- *Defining the costs*: The costs couldn't exceed $100,000.

4.3.2 Best Practice 2: Including Sufficient and Necessary Physical Phenomena

MISTAKE 2

The objective of the study was then defined. Doug was aware of the expected level of accuracy, the life situations to study, and the project result deadlines.

Doug was working on the real and mathematical model of the fridge. How would it be possible to achieve the expected accuracy? Where to start (Figure 4.1)?

The internal temperature of the fridge is a key variable, as it will develop differently depending on whether the fridge is equipped with a fixed or variable compressor, and the compressor will be driven differently (and so consume differently). Doug began the study of the internal temperature.

Then he figured out that this internal temperature is uneven. The closer to the cooling system exchanger the air was, the colder it would be. Thus, the thermostat temperature-measuring tube (activating the cooling system) didn't detect the same temperature as the exchanger. This would impact on the energy use of the fridge.

Doug decided to build a 3D model of the internal air mass of the fridge. After a month of working, he succeeded in building this thermic model of the air mass, taking the geometry of the fridge into account.

Although Doug was proud of this very complex model, it was especially heavy and slow in calculation time. Furthermore, Doug had very little time left to build the rest of the model. He decided to model the cooling system, which he didn't know well, very simply. He assumed the heat energy absorbed by the cooling system exchanger was proportional to the electric energy used by the system.

The complete model was then ready. Still, Doug realized two points. First, his model wouldn't be able to discriminate with enough accuracy ($\pm20\%$ accuracy was expected) the energy use differences between fixed and variable compressor operations. Secondly, the model was too heavy, and no computer was powerful enough to run it.

Thus, the real phenomena Doug included in his model were not appropriate to the study goals, and this led to a dead-end situation (Figure 4.1).

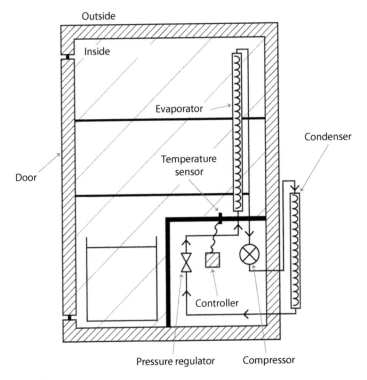

FIGURE 4.1
The studied system: a fridge.

To prevent Doug's mistake, it is vital to step backward right to the beginning, in order to include nothing but the sufficient and necessary real phenomena, to answer the objective of the study. How?

Before choosing the physical phenomena to be taken into account, three prerequisites are necessary.

1. *Identifying the physical phenomena*: This is about building a comprehensive understanding of the intrinsic mechanisms of the system impacting on the observed result variable. We must make sure not to worry about physical phenomena that won't have any impact (this may seem obvious, but it is a frequent mistake made as soon as the topic becomes complex). This study of the inner mechanisms must be team-worked with the project manager or a system expert. Knowing how to decompose the system is necessary for methodical comprehension.

   Let's take the example of the energy consumption of the fridge. What are the physical phenomena impacting on the consumption? An analysis must allow one to decompose the system into three main subsystems: the storage area in the fridge, the cooling loop,

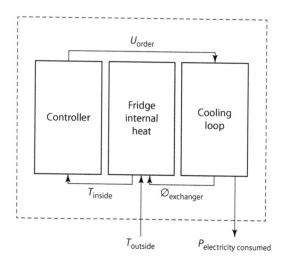

FIGURE 4.2

The three physical phenomena to discern in order to study the fridge's energy consumption.

and the cooling loop controller. So, we notice that the energy consumption will depend on the cooling loop functioning (compressor activation profile), the loop activation will depend on the loop controller activation, and the loop controller will rely on the storage area's internal state (temperature). Three major physical phenomena come into play. Nevertheless, it is still always possible to detail each phenomenon further. Let's take the example of the cooling loop. The electric energy consumption of the compressor will depend on its activation profile, functioning temperature, or even lifetime. If it is always possible to detail more, when should we stop? This is what we'll see next (Figure 4.2).

2. *Ranking the physical phenomena*: This is about distinguishing the phenomena that will dramatically influence the result from those that will not (these are sometimes said to have an order of impact: 1, 2, 3, etc.). There are three alternatives to identification: expertise, live experiments, and simulation. In the previous example, expertise will lead to the conclusion that the compressor activation profile impacts in order 1 on the consumption, while its running temperature impacts in order 2. The reader will notice that these ranking criteria rely on the impact of the results. Indeed, some physical phenomena may seem prevalent, although in the end their repercussions on the final result are minor. As soon as the system becomes complex, the number of physical phenomena to be taken into account increases and ranking becomes arduous to execute. Thus, it is essential not to complexify the model too much or too little. Prioritization is sometimes done retrospectively (we'll come back to this later).

3. *Knowing the required level of accuracy*: If Best Practice 1 has been carried out, this point is immediate. It is just about having the necessary level of accuracy in mind in order to meet the goals of the study.

In theory, after applying these three prerequisites, the simulation engineer knows how to assess the necessary level of physical phenomena to be taken into account in the model. Indeed, he knows the mechanisms at stake, their relative effects, and the required level of accuracy. The sufficient and necessary mechanisms are those to be involved.

In practice, going from one extreme to another is commonplace—that is, complexifying or simplifying too much. We provide some clues here to keep in mind how to find the right balance.

- *Making it as simple as possible*:
  - *The virtues of simplicity*: Creating a model as simple as possible offers many assets. Indeed, a simple model is quicker to develop, is more flexible, and requires less data and power/calculation time. It will be easier to read and maintain. Thus, a simple model is both cheaper and more efficient, while a more complex model isn't valuable.
  - *The limits of simplicity*: On the other hand, a model that is too simple is obviously not relevant. If some of the physical phenomena are not considered, the study objective will not be reached.
  - *The right balance*: It is about making it as simple as possible but not too much, the right balance being defined by the need for accuracy and the study goal. In the end, a system model is often too complex regarding one of the subsystems and too simple regarding another. The objective isn't to design a perfectly fit model but one with the appropriate precision, with a complexity that doesn't weigh on it.
- *Why are models often too complex?* Knowing the reasons will enable one to avoid them:
  - *Reason 1*: It is easier to figure out that a model is too simple than too complex. Indeed, in the case of a model that is too simple, the forecasts appear rapidly wrong (at least more incorrect than expected). On the other hand, an excess of complexity isn't that easy to detect, and we can pass through disadvantages without noticing them.
  - *Reason 2*: Lacking system knowledge. It is often easier to complexify than to properly simplify. A misapprehension of the system may lead to useless complexification.
  - *Reason 3*: Having a taste for complexity. The engineer may sometimes, by inclination, involve greater levels of complexification, even if it is irrelevant.

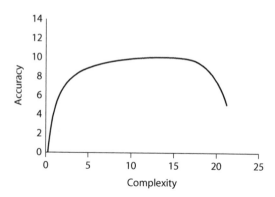

FIGURE 4.3
Enhancing complexity doesn't increase accuracy.

- *Reason 4*: Confusing simplicity with a lack of accuracy. We may think that providing more details enables one to be more accurate. This is sometimes wrong, as mastering fine phenomena is more arduous than a rougher global phenomenon (Figure 4.3). For example, if the purpose is to assess the average temperature of the fridge, a first option is to make a macroscopic heat statement of the air inside the fridge. So, this means modeling the average thermal exchanges of the fridge linings, the food items, and the cooling system. But it is possible to detail more and to build a 3D model of the internal air to highlight the inhomogeneity of the temperature and therefore the unevenness of the heat exchanges, to average the temperature afterward and obtain the expected average value. Still, there is no guarantee a more careful (3D) modeling will provide better results. Setting this 3D model will indeed be more complex, and it is likely that precise data will not be reachable (unlike average data). Thus, in practice, we can often find detailed models less accurate than more global ones (the detail augmentation doesn't enhance it unless it is combined with better expertise).
- *How to simplify*: Two methods can be applied:
  - *Upward method*: The most frequent. The model is initially too simple on purpose to be progressively complexified. The correlation step (see the correlation best practice in Section 4.3.6.2) enables one to highlight the weaknesses or inaccuracies of the model. They are then corrected. This is an efficient but never exhaustive solution, and some errors can't be unveiled this way, but they still impact the final results.

- *Downward method*: The model is initially too complex on purpose. It is then simplified (we often say *reduced*). For example, the study aims will be to assess reactivity to some physical phenomena. Low impacts on the result are then discarded. This method is more accurate but time-consuming.

These recommendations should help with mastering the physical phenomena to be included in the modeling. Before illustrating this with an example, it is useful to mention the restrictions of this approach.

- *Iterations*: It is sometimes difficult to appraise the physical phenomena to reach the expected accuracy. Indeed, if the system is very complex, or if there is a lack of expertise, it may become almost impossible to define in practice, before producing results, the necessary and sufficient mechanisms for the desired level of accuracy. In this case, a common solution will be to freeze the first level of modeling. The next building steps are done (steps 2 and 3 of Section 4.2), and the first results are generated. Then these iterations can be used to draw up a satisfying representation (see the upward and downward methods).
- *Taking into account the technical achievability*: There are often several alternatives to represent the same physical phenomenon, with different approach angles and subtleties. Some representation types will be easier to edit in data (more details in Best Practice 3), and this needs to be considered when choosing the phenomena to represent.
- *Already-existing models*: Sometimes a study is requested and the system model already exists. A part of the approach remains the same: the model needs to be *at least* detailed enough to achieve the study goals (the model may need to be complexified). On the other hand, a model can turn out to be way more complex than necessary. In this case, either the model is unique and is kept, assuming the complexity is unnecessary, or a second, simplified model is built. The option of building a new model will be chosen according to its ability to penalize the study or not (in particular, according to the calculation power/time required, etc.; see the *virtues of simplicity* in this section). In the end, it is common for the same system to have several existing models with different levels of complexity and/or different types of physical phenomena taken into account.

Thus, at this step, the physical phenomena to take into account can be identified, and the system is represented by all of them. This type of representation is called a *conceptual model*.

APPLICATION OF BEST PRACTICE 2

Doug noted his previous mistakes. He complexified the air temperature inside the fridge modeling too much, and he also simplified the
cooling system too much. By stepping back, and after talking with a
technical expert, he decided to model three physical phenomena with
the following sophistication levels:

- *Internal heat energy*: An average temperature will do the job; the
  inhomogeneity of the temperature inside the fridge will only
  have an order-2 impact on consumption.
- *Cooling system*: This requires computing the pressure on every
  point of the cooling loop, in order to discriminate the consumption impact from a fixed compressor to a variable compressor.
  Thus, the effect of the main components of the cooling loop on
  the cooling fluid pressure needs to be modeled. Nevertheless, a
  simple 1D model will do the job.
- *Controller*: The controller logic has already been defined by the
  design teams and is relatively simple, it just needs to be taken
  the same way.

4.3.3 Best Practice 3: Converting into Equations and Configuring the Model

MISTAKE 3

Now that Doug had defined the physical phenomena for his model, he
could turn them into equations and set them. Doug described three
main phenomena; he started with the first, the representation of the
heat energy inside the fridge.

  He decided to only use an average notion of temperature, so he
represented the average fridge temperature T through the following
equation:

$$\frac{mC_m dT}{dt} = \frac{1}{R(T_{out} - T) - \varphi_e}$$

where:

m is the internal mass (composed of the air and the food items in the
      fridge)

C_m is the average specific heat capacity of the internal mass

R is the equivalent heat resistance of the fridge (the plastic and the
      lagging that constitute the outline of the fridge)

T_{out} is the outside temperature
φ_e is the heat power withdrawn by the cooling system exchanger

This last variable, φ_e, will be the outcome data of the cooling system model.

This way, even before going on with the equation of both the other physical phenomena (cooling system and controller), Doug figured out he would need the values of three parameters defining the system: m, C_m, and R.
How could he figure out the values?

He started with the parameter m. After all, this is the internal mass; he could assess the internal air mass thanks to a measurement of the air volume and weigh the items in the fridge (shelf and tray). Then, for the parameter C_m, by identifying the internal materials of the fridge (air, glass shelf, plastic tray), and by figuring out the specific heat capacity of these materials, he managed to build an approximation of C_m. Finally, regarding R, he found data for a similar fridge that seemed to be of the same type and used these values.

Nevertheless, before going further, Doug figured out his assessments were risky. The internal temperature would have an order-1 impact on consumption. With his very approximate assessments, he wouldn't ever be able to reach the 20% accuracy set beforehand. So, his equations and choices of settings have to be reconsidered.

To avoid Doug's falling into this situation of uncertainty, it is necessary to choose the best way to equate and set the physical phenomena to take into account. We are going to review these two points.

4.3.3.1 Converting into an Equation

To depict a physical phenomenon, three types of modeling have to be mentioned and therefore three types of equations.

1. *Physical model*: Fundamental physical laws are used (e.g., the forces of gravity). The phenomenon will be depicted through a range of interactions between several physical laws. The parameters (see Chapter 1) of the model will have a physical dimension.
2. *Nonphysical model*: In this case, the phenomenon is depicted by using an arbitrary model that will have to be set. This can be a strictly linear model, in order to be determined as polynomial, or any other function. The parameters will not have, in this case, any direct physical significance.

3. *Half-physical model*: The two previous notions are not absolutely different, and this arrangement is frequently chosen. For example, a complex physical phenomenon can be modeled in a simple way; we are then moving away from the classical physical laws as they are approximated, without completely losing the physical dimension. Also, a global model can be called *half-physical* if it includes both physical and nonphysical subsystems.

Then, how does one choose among these three potential types of model? The choice has to be made according to the situation's restrictions, considering the following two points:

- *Advantages of physical models*: The first advantage of physical modeling is to reduce (in some cases) the amount of parameters to identify with live experiments (that can be, as we'll see below, an issue). For example, when modeling a phenomenon where gravity is involved, the gravity constant g is already known, making it one less parameter to identify. The second advantage is about the increase of the model's reliability. A model stemming from the laws of physics is more often closer to reality than a mechanically configured model. Finally, the last advantage is an easier reading of the results, enabled by the physical interpretation.

- *Disadvantages of the physical models*: The main disadvantage of this type of modeling is a costlier development time. Indeed, when the system is misapprehended, or very complex, it may take time to describe the phenomena involved then combine them and simplify them. In comparison, the definition of a nonphysical model is done almost instantly. Also, in some cases, the definition process may reveal a plethora of settings that individually have a very low impact on the final result. In this case, spending time identifying each of these parameters may not be the best strategy.

  Now, how does one define the best equation for the physical phenomenon?

- *Physical model*: The equation will rely on the system concerned. This topic will not be detailed here, and the reader should refer to specialized books if needed. Then it is about establishing a sharp understanding of the physical laws associated with the problem. Many types of software can noticeably help the simulation engineer during this process (see Best Practice 4). The equation time can be decreased or even cut, while the simulation engineer carefully keeps on stepping back, bearing in mind the limits of modeling.

- *Nonphysical model*: This case involves deciding on the type of universal model to use (such as the order of a polynomial). We will not

cover this topic either as software is becoming more and more helpful to the simulation engineer, and specialized books have already
been written about that. Let's just mention a good commonsense
practice: always start too simple and then complexify.

4.3.3.2 Configuring

Once the simulation engineer has finished the equations, the configuration
of the model is still to be done (the identification step).

The parameters to identify are inner constants characterizing the system,
the initial instant state value variables, or even limit conditions. The parameters identification step is especially delicate, as it strongly impacts on the
model's final accuracy and it may turn out to be very complex. In general,
the identification process is performed once the model has been coded (see
the following best practices). We decided to review this step at the same time
as the equation, because it is related and often spurs one on to review the
way the physical phenomena are equated. We'll touch upon how to rele-
vantly configure the model.

Before the identification step, a few prerequisites:

- *Choosing an identification method*: There are two alternatives:
  - *Direct method*: The approach to favor if possible. It concerns iso-
    lating a parameter and measuring it directly (or finding its value
    from known physical constants). For example, the fridge mass
    is easy to find with direct measurement, while the heat capacity
    isn't. The direct method mainly deals with physical models, but
    it is not always possible.
  - *Indirect method*: The one to use if the previous one isn't possible.
    Parameters stay to be identified. The simulation engineer uses
    live experiment results on the system and identifies the remain-
    ing parameters. He uses optimization techniques to minimize
    mistakes. We'll review this technique in detail below.
- *Anticipating to reduce work*: Some precautions must be taken to ease
  the identification work:
  - *Choosing the way to represent a physical phenomenon*: We men-
    tioned this point in Best Practice 2. Indeed, several alternatives
    often represent the same physical phenomenon, and the param-
    eters to be defined will then be different. Some parameters are
    easier to determine than others. It is important to anticipate
    this point during the previous phase to simplify the identifica-
    tion work.
  - *Decreasing the amount of parameters*: Unknown parameters are
    often gathered in equations in such a way that it becomes possi-
    ble to combine them and reduce the parameters to be identified.

Let's illustrate this through a very simple example. Assume a force F on a system may be written as $F = (A + B)\,x$, where x is its position and A and B are two parameters to determine. If these two parameters don't appear in some other equations, all there is to do is to define a new parameter α, in the way that $\alpha = (A + B)$. Thus, only one parameter is to be identified instead of two. This point may seem obvious, but as soon as the equations become complex, discerning the possible gatherings may require analysis work.

- *Assessing the parameter-related accuracy impact*: In order to respect the model's accuracy goal (Best Practice 1), especially in terms of mastering the accuracy of the parameters. It is sometimes possible to set the accuracy of a parameter. Other times, it is more complex. A common method would be to assess the parameter's sensitivity and its impact on the final result. Thus, it is possible to know which parameters are the most weighty and so which are to be characterized with the most precaution.

Once the three prerequisites have been applied, the true identification work may start.

In the case of direct method identification, the task is less complex, or at least the difficulty is essentially in the knowledge of the studied system and its measurement techniques. Thus, we'll lead the reader to specialized technical books. On the other hand, in the case of indirect method identification, the task is critical, and the difficulties are often generic and not related to the system concerned. We'll thoroughly cover this topic, and we're going to detail the best practices to ideally use in this indirect method.

- *Knowing the principle of identification*: As we mentioned before, this deals with defining the model's parameters by reducing the gaps between the model and live experiment results. Let's illustrate this through a very simple example. Assume we have modeled the speeds of a system with $s = a + b\,t$, where t is the time. We then decide to perform a live experiment on the system, measuring its speed from time $t = 0$. The results are shown in Figure 4.4. Then, the simulation engineer identifies the parameters a and b as the modeling result $s = a + b\,t$ as precisely as possible. In other words, it is about minimizing the gaps between the obtained model and the measurement. There are different methods, such as least squares, consisting in reducing the sum of squared gaps for each measurement point. Through this least-squares method we get the following identification:

$$\begin{cases} a = 2.8 \\ b = 0.52 \end{cases}$$

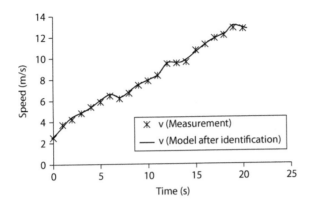

FIGURE 4.4
Identifying the model parameters to minimize the simulation/test gaps.

In practice, the identification is often tougher, because it is a more complex model, there are more parameters, and the parameters are less easy to master.

- *Taking advantage of software contribution*: Nowadays, a lot of simulation software assists the simulation engineer in the parameter identification process. Thus, it is less common to have to master the mathematical optimization methods for the best parameter identification. It remains decisive to be able to define the measurements in order to complete the identification phase and to know about the usual traps and limits. We'll review these specific points.

- *Defining measurements to make*: Or choosing the measurements on which the parameter identification will be performed. First, it is about making sure the signals to be assessed are as sensitive as possible to the parameters to be measured. The best method to put a strain on the system (inputs choice) and the variables to measure must be chosen. Also, the measurement volumetry must rely on the need for accuracy and on the multiplicity of the phenomena involved. The tests must be done in the closest conditions to those in which the system will operate. Finally, the measurement sampling must be consistent with the time constants of the phenomena parameters.

- *Making the best use of the measurement results*: A measurement result must be considered with as much caution as simulation. The measurements issued in the live experiment are also an approximation of reality. Post-treating raw results is often required before using them for parameter identification. It is then about putting aside the results we know not to be representative, adjusting values where there is an offset, and even filtering the data to avoid identification problems. Data frequently have to be resampled.

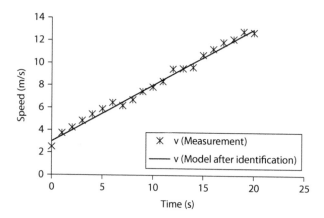

FIGURE 4.5

The overfit trap: the model approaching the tests, although this is irrelevant.

- *Avoiding the overfit trap*: With the best intentions, the model may be expected to approach the measurements taken "too much" (also called *fitting*). This may be counterproductive. Indeed, for a given set of measurements, it is always possible to create a model that perfectly fits the data. All we need is to complexify the model and increase the amount of parameters (especially in the case of a non-physical model). For all that, if a new set of data is produced with the same system, the model created is very likely to be less reliable in predicting it. Let's go back to the previous example, in which we modeled the speed through $s = a + b\,t$. Trying too hard to fit the measurements may lead to complexifying the model and thus to adding new parameters to obtain, for example (reproduced in Figure 4.5):

$$s = a + bt + ct2 + dt3 + et4 + ft5 + gt6 + ht7$$

Of course, this new model measurement is more correct than the previous one, but it is very likely to be less accurate if a new set of data is created. Thus, an overfit model may become nothing but a measurement copy and lose any physical dimension, including its ability to generate a new set of data.

To prevent overfitting, we mustn't want to fit with measurements. It may also be wise to only use the part of the measurements available for identification and then monitor the ability of the model to predict measurements that were not used. In the previous example, we could use measurements going from $t = 0$ s up to 15 s for identification, and use measurements from $t = 15$ s up to 20 s to check the model's relevance. This method could help figure out that the overfit model was in the end less accurate than the simple one.

- *Knowing the limits of identification*:
  - *Impossibility of identifying some parameters*: In some cases, some parameters can't be identified (through the indirect method). If we go back to the example with the model $F = (A + B)\ x$, identifying the parameter $\alpha = (A + B)$ is achievable, but separately identifying parameters A and B is not. Also, regulators may compensate some effects, and in this way make the identification of some parameters impossible.
  - *Impossibility of performing representative measurements*: Experimenting and generating expected inputs or measuring the variables may not always be possible. Also, the properties of the system may evolve over time, making any identification obsolete.

When these limits are encountered, it is not about giving up the study. Indeed, several solutions are often possible and have been detailed previously: modeling the system another way, approximating the parameter, or using a direct method to assess it.

APPLICATION OF BEST PRACTICE 3

Doug noticed his parameter identification work was hazardous and got back to work.

He started with deciding to keep the equation he defined, which is

$$\frac{mC_m dT}{dt} = 1R(T_{out} - T) - \varphi_e$$

Indeed, on the one hand, this type of equation has a physical significance that would enable better understanding of the model and better accuracy than with a nonphysical model (e.g., a polynomial type). On the other hand, this model would be quick to use as it uses simplified notions (e.g., average heat capacity, equivalent resistance).

Doug noticed he could perform tests on the fridge to identify the parameters. Indeed, besides the cooling system, the new fridge would be the same as those from the previous generation. The tests to perform remained to be determined.

Doug also remarked he could easily go from three to only two parameters by assuming $C = mC_m$. He rewrote the previous equation this way:

$$\frac{dT}{dt} = A(T_{out} - T) - B\varphi_e$$

where A and B are the two parameters to identify. Thus, Doug noted that both these parameters matched with two different physical effects. The coefficient A described the internal/external heat exchange impact on the temperature inside the fridge, whereas the coefficient B described the heat power withdrawn by the exchanger-related impact on the inner temperature.

This understanding enabled a better definition of the test to identify these parameters. It was fitting to start measuring with a cold fridge and cut any power supply. The inner temperature would increase, according to parameter A. Then, after a chosen time, the supply was restored, and the temperature decrease dynamic would be under the influence of parameter B. This experiment would enable an accurate identification.

In the end, Doug measured the outside temperature at around 20°C, and he restored power after 5 h, with a 50 W estimated flow in the exchanger. He measured the temperature every 15 min, and got the results shown in Figure 4.5.

He just had to identify A and B by minimizing the model's mistakes. He used some of the widely available software and got the following as a result:

$$\begin{cases} A = 5.6 \ 10^{-5}\,\mathrm{s}^{-1} \\ B = 9.7 \ 10^{-5}\,\mathrm{KJ}^{-1} \end{cases}$$

Doug successfully put into the equation the first physical phenomenon (the internal heat of the fridge) and parameterized the equation in a pertinent way. He now has to do the same with the cooling system and the controller (we will not this describe here; Figure 4.6).

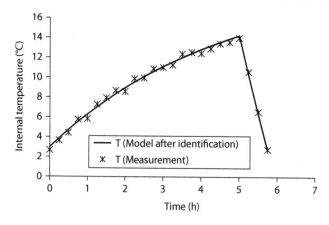

FIGURE 4.6
Model parameter identification with test results.

4.3.4 Best Practice 4: Picking the Software

MISTAKE 4

After making it an equation and defining the model's parameters, Doug had to implement it into software so that his computer would produce the simulation results. The only software known to Doug was SinuM. (This name has been invented for the example; we are not advertising any developer here!) He decided to use this tool to implement the equations.

So he coded the fridge's internal heat equations, as well as those of the cooling system, using SinuM. He realized he didn't have to build a model for the controller. Indeed, the model of the controller had already been built for its development. However, the controller was modeled with SinuT, which isn't compatible with SinuM.

Doug didn't want to entirely recode the controller code in SinuM, so he had to reconsider his software choice. His coding time spent in SinuM was wasted.

A little frustrated, Doug realized he could have prevented this waste of time by picking the appropriate software.

In order to avoid Doug's mistake, the software to implement the model must be chosen wisely and as early as possible. Two specific cases often occur.

- *The question doesn't really arise*: Often, the simulation study is an extension of previous ones. In this case, the chosen software is likely to be the historical one (even if, of course, new constraints or opportunities that may occur must be identified). Indeed, changing software or integrating new software isn't trivial, and this must be done only if necessary or contributive.
- *The question arises*: This is the case we are about to develop in more depth in this chapter. As soon as a topic makes a break with previous works, it is relevant to ask this question.

Nowadays, there are many types of software on the market, each with pros and cons. It is then appropriate to choose the best one and cross-check the following:

- *The need*: What are the main features that must be fulfilled by the software? What are the company restrictions?
- *The offer*: What features of the software are available on the market?

We are going to review the five main factors to take into account when opting for simulation software. For each of them, the reader will be able to identify the need, whether the considered software answers it, and then rank this factor's stake among the others. Then a wise choice can be made. The five main factors to be taken into account are the following.

4.3.4.1 Coding and Model Design

- *What type of model does the software fit?* The types of models were explained in Section 4.3. No software today can really multitask. For instance, very few software programs are efficient at both 1D and 3D models.

- *What type of physics does the software fit?* Some software programs are more efficient at performing mechanical studies, while some others are more designed for digital ones. The editors intend to reduce this factor, but there are still imbalances.

- *What ergonomics/support is there in the modeling?* Some software programs are very raw and close to the coding language, while some others support the simulation engineer by only giving them the task of building the conceptual model (in this case, even the physical equations are made by the software).

- *What compatibility with other software does it have?* Some software can easily be interfaced with competing editing software, and it is convenient for the simulation engineer. Also, this opportunity may be more or less ergonomic.

- *What additional packages are there?* With the amount and diversity of the packages, the only limit is the editor's creativity. Some features can be very useful to the simulation engineer, such as the growing libraries, helping them to use previously built standard submodels.

4.3.4.2 Model Simulation and Analysis

- *What is the performance?* The calculation speed and the accuracy of the results for a single model can vary according to the software used. This may become a decision criterion by itself if real-time objectives are required. Indeed, in cases where human interaction is needed (e.g., flight simulation), being able to simulate in real time becomes a must.

- *What results display is there available?* Software is allowing more and more ways to display the results (e.g., in 3D), with several features. This may become very useful for a better understanding of the system or for broadcasting the results.

- *What post-treatment usability and features are there?* After making the simulations, it is usually necessary to arrange the obtained data, even making changes (filtering, etc.) The software may support the simulation engineer in this task.

- *What additional packages are there?* As for the modeling process, there are many packages available for the simulation process. For example, many software programs help the model parameter identification step or enable one to achieve optimization campaigns.

4.3.4.3 Support

- *What level of internal support is there?* If the software is already known by the company, this is an additional reason to choose it. Indeed, many company users can train and assist new users.
- *What level of editor support is there?* Some editors are known for having very efficient technical support. This may be crucial, especially when the software is complex or if some features must be specifically developed for the customer company.
- *What other support is there?* There are some other tools to assist the simulation engineer when needed, such as editor-written documentation or Internet forums. All software programs are not equal regarding the quantity and the quality of information available.

4.3.4.4 Actual Costs

- *What are the costs for the license?* This is often the first software cost item, but its definition must be made carefully. Is the license bought or rented? Is it related to a specific computer, user, or group of users? Are there several types of licenses regarding the type of package used? Are software updates included?

  So, during the comparison of the different software license costs, what can be compared must be, acknowledging many clauses exist. So, comparing software license costs to compare what is suitable for comparison, acknowledging many current clauses.

- *What direct costs are there, excluding licenses?* The money charged by the editor, for example, for technical support.
- *What indirect costs are there?* Two indirect factors must be considered in the financial analysis, such as the simulation engineer's in-house training period, varying with the software ergonomics.

4.3.4.5 Long-Term Sustainability

- *What coherence is there in the software portfolio?* Companies must streamline their own wide range of simulation software and make sure software programs supply each other's lack. Managers consider

62 *Modeling and Simulation*

side effects when software is to be replaced (including training and adapting former tools to the new software).

- *What is the long-term editor/software reliability?* A brand-new custom-made software program from a small editing business (or even developed internally) may look attractive, but long-term studies require one to observe the reliability of that solution.

- *How professional is the editor?* As for the quality of technical support, the supplier's efficiency in good client service is unequalled, as well as in software maintenance and updates.

- *What software is expanding outside the company?* The more widespread the software is, the more it becomes a standard, and in this way, interesting in the long term. For example, if some models must be exchanged with suppliers, it is convenient for these suppliers to use the same software.

These five factors are a first-reading checklist. They are not exhaustive (especially as constant software upgrades lead to new differentiating factors). The reader will notice that software mustn't be considered simply good or bad but rather adapted or not to the study and the company. Finally, to choose software, the software choice best option analysis will be more or less intensive. If this is a major decision, it will be appropriate to supplement the analysis with commonsense practices such as referring to experts' and users' opinions or requesting a trial software license before purchasing it.

APPLICATION OF BEST PRACTICE 4

Doug went back to his software choice and used the previously mentioned reading grid.

He started by defining his need in terms of software—that is, detailing his expectations and ranking them (refer to Chart 4.2). Then, after excluding many software programs, he compared three considered on and SinuB.

In the end, he kept SinuT, as it is compatible with the controller model (already coded under SinuT). SinuB (as illustrated in Chart 4.2) is also compatible with SinuT, but the technical support is very weak, and that was an important point for Doug. Finally, it appeared that SinuT computing time performances were lower than expected, but he considered this expectation to be insignificant compared with the others.

For all these reasons, he decided to code his whole model in SinuT.

Features	Need		Offer			
	Expectations	Weight	SinuM	SinuT	SinuB	
Encryption and model design	Building a multiphysical 0D/1D model, suitable for SinuT	+++	Ok, except for suitability with SinuT	OK	OK	
Model simulation and analysis	Simulating faster than in real time, parameters identification package available	+	OK	ok, except for weak calculation time performances	OK	
Technical support	Efficient support (because beginner)	+++	OK		No support, new software	
Long-term viability	No specific need	+	OK			
Actual costs	Less than $40k for the whole study	+	OK			
Choice				OK		

CHART 4.2
Reading grid used to identify the most relevant software offer.

4.3.5 Best Practice 5: Managing the Numerical and IT Issues

MISTAKE 5

Doug chose the software he wanted to use and mathematically defined his model. Then it was about coding it into the software, to perform simulations.

So, Doug proceeded with this meticulous work, consisting in translating each equation in SinuT, the previously chosen software. Forecasting to digitally define the life situation to assess must also be done, which is the test environment and the system stresses. After a few days of work, the numerical model and the life situation were completed. All that was left was to test the model!

Doug, with a mix of enthusiasm and worry, clicked on the "run" button to launch the simulation.

After a few more seconds of suspense, a message unfortunately appeared: "An error has occurred." A short time after, the software shut down, without further information.

A little tense, but persistent, Doug repeated his attempt several times.

He opened the software again, ran the simulation again, but always got the same result. Where did the error come from? He had no idea; his digital conversion work seemed perfect to him.

So, Doug had a relevant mathematical model, but he couldn't sort out his digital and IT issues.

We will define here how to best control digital and computer issues, in order not to find ourselves in the same situation as Doug. Two issues must be identified.

1. When considering a real system, how far can a mathematical model properly represent it?
2. When considering a mathematical model, how relevant can the results provided by a numerical model be?

The objective of this section is only to study this second issue (we have already started studying the first one and we will get back to it in the upcoming best practices).

The ability to properly develop a numerical model from a mathematical model depends on the deep knowledge of the software used, which is not the purpose of this book. Nevertheless, some numerical/computer issues are generic, and it is relevant to approach these here.

So, we will proceed in three steps. First, we will clarify these types of issues, then see how to prevent them, and finally see how to deal with them if they happen.

4.3.5.1 What Are the Main Types of Numerical/IT Issues?

- Simulation doesn't provide a result: This is the situation Doug met; we talk here about a *bug* or *crash* in the model. This issue has the advantage of being visible.
- The simulation provides a result, but
  - *The calculation speed is too slow*: The simulation works, but the time to obtain the results is too long and may be prohibitive.

- *The simulation results are wrong*: With a warning as seen previously, we have a case where the simulation results are noncompliant with the mathematical model (so the lack of accuracy related to bad mathematical modeling is excluded). For example, if Doug simulated the electric consumption of the fridge in 1 day and got 0 Wh, this could come from a numerical issue, not from a mathematical modeling issue. In the end, this kind of issue is often the most difficult to deal with because it is difficult to identify (especially if the result seems likely). It is better not to obtain any result than a wrong one.

4.3.5.2 How to Avoid Numerical/IT Issues

The proverb "Prevention is better than cure" makes perfect sense here. The few hours of work invested to build a good model allows one to save the many days that would be necessary to solve such an issue. So, it is essential to comply with the following:

- *Know the common issues*: This allows one to anticipate them and to adapt the numerical model. This is the topic of this whole section.
- *Structure the model*: This is a staple, sometimes a little boring, but extremely profitable as soon as the model becomes complex. It then deals with
  - *Defining blocks*: There are often several ways to define a shape, the point being that there is one. It is especially possible to gather from subsystems or according to the physics at stake. Regarding Doug's example, he needs *a minima* to structure the model into three main blocks: the fridge's internal heat, the cooling loop, and the controller. Then the sub-blocks can be defined.
  - *Dissociating the documentation and the model*: It is frequent (and recommended) to write comments on the models, which won't be read by the software solver but by the simulation engineer (or his colleagues). This documentation must also be structured and dissociated from the code.
  - *Taking the constraints into account*: Several constraints must be considered to structure the model. On the one hand, the model structure must allow one to understand it quickly. On the other hand, it must be flexible to enable the easy adjustment, withdrawal, or addition of elements. It may also have an impact on the calculation speed and performance. Lastly, the structure must enable the simulation engineer to be efficient in his coding.
- *Perform checks*:
  - *During the construction*: It is fitting to test the model as it is being built. This way, each time a model block is completed, it is

appropriate to test it (by defining the constants at the input of the block and checking the output). If the test is conclusive, the block is approved and can be connected to the previous one. This way, if a problem occurs, its identification will be much easier than if it is made once the global model is assembled.

- *Before some specific simulations*: Some simulations are specifically long, and may last several days or months. Certainly, any numerical engineer will have already encountered the situation of being forced to wait several days for a result that was already wrong after a few seconds of calculation. So it is wise, before running a long simulation, to perform tests on a reduced situation, to correct possible errors before starting the entire simulation. Also, if batches of simulations are made (an automatic launch of a set of situations), it is wise to check with a reduced batch.

- *Systematically report errors*: It is pertinent to include in the code mechanisms allowing one to warn the simulation engineer if any error is detected. This not only permits one to prevent a mistake from being included in the simulation (which is not necessarily easy to detect otherwise) but also helps to quickly identify and isolate the problem if the model is complex.

4.3.5.3 How to Solve Numerical/IT Issues

We saw in the previous section how to avoid making mistakes as much as possible. Even if many errors will be prevented this way, some of them will have to be dealt with. We provide here the main tools to achieve that.

Identify and isolate the problem:

- *Stake*: This first step often represents 90% of the work to do to solve the problem. Indeed, a great amount of issues are, in the end, very simple to fix once they are isolated.

- *Method*: Two specific cases are possible. Either the simulation engineer, or any other person involved, has an intuition about the origin of the problem, in which case it is relevant to immediately investigate on that intuition. Or the simulation engineer has no clue, and a structured approach must be adopted. This approach may be conducted by determining whether the problem is numerical or IT related (see the upcoming section). Then, it may be appropriate to decompose the model into blocks, in order to test each block separately and identify which one is defective, and reiterate this process further and further into the model. This systematical process can be very time-consuming.

- *Snowball effect*: One of the main obstacles to identifying the problem comes from a mistake often provoking another one. Thus, even if it is not always easy, one must know how to identify the symptoms of the root cause. To illustrate this in a simple way, let's assume that in the model an internal variable value is, by mistake, 0. If, further into the model, a second variable is the result of a division by this first variable, the result will be infinite, which will provoke a mistake. The error comes from the first variable, not from the second.

Fixing the problem: In the end, many numerical/IT issues are very frequent, and we provide a few examples here.

- *Computer issue*: In this case, the numerical model may be valid and still cause computer issues.
    - *Software and hardware compatibility*: For example, the software may not work with some hardware versions.
    - *Model compatibility with the software*: For example, the model has been designed on a former version of the software, and the new one does not allow use of the model as it is.
    - *Model environment*: For example, the model requires access to some databases or other models that are not available.
    - *Other computer restraints*: For example, the file name including the model can't involve special characters, or the entire address can't involve more than a number of characters.
- *Numerical issues*: In this case, the computer environment may not be valid; thus, the model may cause numerical issues.
    - *Code incomplete or not depicting the mathematical model*: This may simply come from a human mistake: an omission in the code or some written error (because of careless mistakes or software misreading). These common mistakes are difficult to avoid when the code becomes long and complex (even more if it is developed by several persons). This is why the code has to be monitored progressively (see the next paragraph).
    - *Nonsimulatable code*: For example, some variables used in the code aren't defined, or some equations defined in the code are irresolvable.
    - *Wrong time (or space) discretization*: As seen in Chapter 1, many models require time or space discretization, and the intervals must be set by the user (potentially in the solver parameters of the software). Obviously, a discretization that is too wide may lead to inaccurate results, while one that is too fine may blow out the calculation time.
    - *Impact of rounding and/or truncating*: No matter how powerful computers are, they aren't infinitely accurate: they round and

truncate, leading to potential mistakes—for example, if the code sequences these two operations:

$$x = 10 - 20$$

$$y = 10 - 20x$$

Depending on the rounding management, x can be approximated to 0, so the calculation of y will correspond to a division by 0, which generally provokes errors.

- *Convergences management*: Many codes use iterative loops, programmed to interrupt only when some requirements are fulfilled. If these are too restrictive, or if the coding isn't efficient enough, these iterative loops may be very long or even endless.

- *Discontinuities management*: Discontinuities are often a factor of numerical problems. Indeed, many software programs mishandle discontinuities and may interrupt the calculation (e.g., if the discontinuity causes an endless slope) or dramatically slow it down (if the calculation time discretization must be increased to filter discontinuities). The inception of the simulation may be a discontinuous event, inducing a careful definition of the system's initial variables.

- *Software bug*: Editors are not foolproof, and software can have some anomalies, which can be expressed when certain factors are combined.

So, we have covered how to avoid numerical/computer mistakes and how to fix them when they still occur. Let's mention a few more points before moving to the next best practice.

- No result is better than errors. We have already mentioned it, so let's get back to this point in the enlightenment of the last section. A set of factors may help in getting a result but unfortunately a wrong one (which is worse because wrong decisions will be made from the results). That is why numerical model results have to be put into perspective.

- If the numerical model is perfect, results will be as good (or as bad) as the mathematical model.

- The identification and error solving process is more and more software aided.

- Of course, software completes what it is asked to. In the end, errors always come from human mistakes: being aware of the tool's limits is crucial.

- Software developers are the most capable of identifying model errors. If the user is not the creator, he or she must get in touch with him or her if necessary.

- The numerical modeling process will be much more effective if the expert network and technical support work within the company. This will be covered in Chapter 5.

APPLICATION OF BEST PRACTICE 5

Doug became aware of his mistakes and reviewed his numerical modeling work. Since he had defined three main subsystems (internal heat, cooling loop, controller), he modeled and checked them one by one.

He started with the internal heat model. After modeling it, he decided to test it by freezing the variable income data, which will then be coupled to both other subsystems. There, the simulation works and the results seemed credible.

He moved to the next subsystem, the cooling loop; and proceeded the same way. This time, the simulation didn't work, and as before, the software ended by shutting down without any further explanation. Doug had properly decomposed the same subsystem into five subsystems. Again, he tested each of them separately. Only one was a problem.

Now the situation is easier to analyze; the code of that subsystem is only a few lines. He then figured out the mistake. Doug used an iterative loop, without defining the convergence condition. So the calculation never stopped. He corrected that simple mistake, and henceforth the numerical model of the cold loop worked.

Doug carried out his work to numerically model the whole fridge.

4.3.6 Best Practice 6: Managing the Validity Level of the Results

MISTAKE 6

After weeks of work, Doug managed to build a numerical model of the fridge studied. Also, he managed to get some simulation results with the model, to check it was operational.

The study's purpose was to predict, regarding a specific situation, the consumption of the new type of fridge (which had a variable compressor instead of a constant one). We saw that the expected accuracy was ±20%. Doug announced his results to the project manager.

Doug: "The model predicts the new fridge consumption will be around 50 W given the life situation initially defined."

> *Project manager:* "Very good, but what is the accuracy of this statement?
>   Is it within ±20%?"
> *Doug:* "I think so."
> *Project manager:* "Are you sure?"
>
>   Unfortunately, Doug was unable to answer. His objective while build-
> ing the whole model was to reach this accuracy, but did he achieve that?
> He knew that if he couldn't tell, his result would add little value to the
> project. Thus, Doug was in a dead-end situation, as he didn't compre-
> hend the validity of his results.

We are going to cover Doug's model validation issue in this section, to avoid
the situation he encountered.

First of all, this matter is extremely important. A result is useless if it is not
combined with a high level of reliability. In the worst case, it will be associated with a randomly generated result. So, results accuracy can be requested
legitimately.

Secondly, this topic is highly complex. It is often easier to build a model
than to manage its accuracy level. On the one hand, as we were reminded
earlier, the model is nothing but a depiction of reality and will never be
perfectly accurate. But on the other hand, the model includes a specific
number of physical phenomena defining the system behavior; there is
always some truth in the results. In the end, between these two extremes,
how do we assess the validity level of the model? That is the purpose of
this section.

Let's start by defining two notions:

1. *Validation*: A simulation result is considered valid if the gap with
   reality is smaller than the previously established maximum gap.
   Regarding Doug's example, he declared 50 W the result, with the
   objective of being ±20% accurate, which means he expects the actual
   consumption to be between 40 and 60 W. The simulation is only
   valid if the consumption is between these values. The notion of validation then corresponds to the modeler's work consisting in verify-
   ing that the results are valid.

2. *Validity domain*: The validity domain is the set of life situations in
   which the model must provide valid results, the life situations being,
   as mentioned before, the definition of the system stresses and the
   environment definition. Obviously, a model's validity domain is
   always restricted to specific life situations. The awareness of the
   validity domain, defined in the study objective, must determine the
   model's construction work.

These notions being defined, we can state two things.

1. First, in theory, we have already detailed in the previous chapters how to make the built model valid in the domain in consideration. Indeed, that was shown sequentially:
   - Best Practice 1 allowed us to see how to define the study objectives, including defining under what conditions the simulation result would be considered valid or not.
   - Best Practice 2 showed us how to identify the physical phenomena to be taken into account, to make the conceptual model valid, starting from the actual system.
   - Best Practice 3 highlighted the making of the model's equation and the parameter definitions, to validate a mathematical model, starting from the conceptual model.
   - Best Practices 4 and 5 allowed us to see how to handle numerical and IT issues with a valid model, starting from the mathematical model (this step is sometimes called the *verification phase*, not to be confused with the *validation phase*, which is more general).

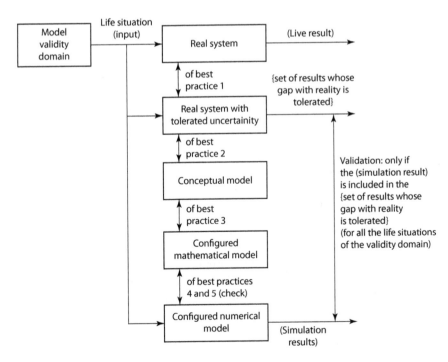

FIGURE 4.7
Model validation: sequential method and general method.

So, if these steps are properly completed, the numerical model has to be valid. This is depicted in Figure 4.7.

2. Nevertheless, in practice, it may be difficult, even impossible, to achieve this work upstream and sequentially. That is why we are going to now detail the additional actions to perform in order to confirm the model's validity or not.

How do we validate the model? There are two possible methods.

4.3.6.1 Sequential Method

- *Upstream work*: The activities to be performed as soon as possible in case some details need to be adjusted during the model designing process.

- *Final work*: The sequential method can be applied again, in more hindsight, once the numerical model is finalized. Asking for external opinions is always appropriate, especially from numerical experts and systems-aware project stakeholders. Each step of a valid model can be analyzed in hindsight by monitoring, for instance, the following:

  - Do the required simulation accuracy levels and the variables answer the client's project needs?

  - Were all important physical phenomena really taken into account?

  - Do the equations and configurations selected enable sufficient model accuracy?

  - Does the model's conversion into numerical language imply a loss of accuracy, making the model invalid?

4.3.6.2 Global Method

This second method consists in testing the model's validity as a whole or straightforwardly testing the building process result: the numerical model. There are three options:

1. *Comparing the numerical model with the real system (correlation)*: This validation method is certainly the most popular and inescapable. It simply consists in comparing the numerical simulation result with the result obtained from a live experiment. This allows one to see to what extent the model is valid. Here are some key highlights for correlation:

   a. *Defining the measurements to perform*: First, this depends on the need. It is indeed wiser to test life situations that could reveal the model's shortcomings—the most important points to check. Second, this depends on what is achievable. Many measurements of less relevant life situations are available. It may still be convenient to use them, even with less benefit.

b. *Using the test results*: See Best Practice 3. Tests results must be considered as cautiously as simulation results.

c. *Not confusing identification and validation*: The aim is to make sure the model is valid regarding the considered life situation. When the model is declared invalid, there is a temptation to change some model parameters and make it valid. Two possible cases about this tricky matter are as follows:

    i. The model is invalid as it was incorrectly set up. In this case, the parameters can be changed (the identification step covered in Best Practice 3). The live experiment may be used to proceed to this setup, but then it is no longer considered a validation test.

    ii. The invalidity of the model is not attributable to the parameters: something else has caused the problem (are any physical phenomena taken into account? etc.). It is essential not to "cheat" by changing the parameters at any cost to get the illusion of validity.

To judge this specific case, it is about only using one live experiment for validation, without using it to identify the parameters.

d. *The limits of correlation*:

    i. *Non-exhaustiveness*: A simulation/live experiment comparison is only performed in a few specific life situations. Thus, it doesn't ensure the model's validity in the whole domain of validity.

    ii. *Good results, wrong cause*: Some simulation results happen to be very close to live experiment results, even if the model considered is invalid. As soon as the model becomes complex, errors may balance each other out, producing a correct result. By chance, this may come up with a life situation tested in correlation, but it is very unlikely to do so in the life situation that will be used for the study.

    iii. *Impossibility of representative measurements*: We mentioned this point in Best Practice 3 (identification), which remains applicable for the correlation work.

2. *Comparing the numerical model to another one*:

a. *Why compare with a model?* This method is often used when the last method limits are reached and when there is another system model. The model is often compared with a more complex and accurate model but can't be used for certain reasons (e.g., calculation time). The model is sometimes compared with a very simple one, to check the orders of magnitude.

    b.   *The limits of this method*: Pretty obviously, this method is only relevant if the reference model has a sufficient validity level.

3. *Comparing numerical model results with expert opinions*: These opinions must be collected not only on the life situation used for the study, which is sometimes too complex, but also on simpler life situations, and the expert will be able to validate some behaviors or not. Several experts in different domains may be involved, validating all the aspects of the model.

    Thus, we have covered the main means to monitor the validity of the model. We mention here only a few observations before reviewing the practical application by Doug.

- *Not overestimating correlation compared with sequential methods*: The correlation method is widely considered more important than it actually is. It is often overestimated because it only validates some specific life situations. On the other hand, sequential methods are often underestimated. Some simulation engineers hope to get valid results even though they have forgotten the major physical phenomena of the system's behavior. Comparisons with live experiments often reinsure the simulation project's customers. If it is true that valid models are those that are correlated, this is mostly because valid models are designed by modelers who care about the validity of their model: they don't overlook correlation. In other words, even if it is less visible, the sequential method is extremely important.
- *Validation and iterations*: The validation phase often involves iterations to modify some aspects of the model (conceptual, mathematical, or numerical). This is not an issue, but anticipating this work is always profitable.
- *Validity domain*: Both reverse traps are frequent:
    - *Using the model out of its validity domain*: Purposefully or not, appealing to the model outside its validity domain is eventually very frequent. Outside this domain, the model has, theoretically, no more value. A great amount of errors ensue from this use.
    - *Not using the model out of its validity domain*: Conversely, it may be too conservative not to try experiencing the model out of its domain, although a project may require it. It is then relevant to do so, without losing sight of the model's potential validity loss.

In the end, modelers are often too confident about the accuracy level of a model used within its validity domain and have too pessimistic a vision of the representativity of the model outside its validity domain.

- *Virtues of open models*: An open model allows any user to go over the decode details (without necessarily being able to change it). On the other hand, a closed model keeps all or part of its code hidden to the user (potentially for confidentiality reasons). Essentially, validating a model is easier in a wholly open configuration. Indeed, this enables users and experts to brush up on the model and its various interactions. We will cover more of these qualities in Chapter 5, Section 5.2, dedicated to organizational best practices (Best Practice 7).

- *Limits of validation*: For all the reasons mentioned before, it may be impossible to perfectly validate a model (even with an acknowledged level of uncertainty). A pragmatic and critically enlightened mind is most important to figure out if the model's use will properly lead to correct decision-making. However, any decision-making tool has its own uncertainty. Neither experiences, experts, nor simulations are perfect. When the storm rumbles, simulation may be limited, but it sometimes remains the only shelter one can find; but one must be aware of its validity level.

APPLICATION OF BEST PRACTICE 6

Doug took note of the best practices to handle his model's validity level and thought again about his response. First, he noticed he properly applied the sequential method during the model's construction, and knew that, hypothetically, the model base was sound. Then, he asked an expert for his opinion on his model. The expert offered to change some settings of the calculation solver he used, in order to get more precise results—that is, a numerical model closer to the mathematical version. Then he decided to compare his model with a real system.

The new fridge did not yet exist; only the former version was available for testing. He decided to upgrade his model to be able to set it the same way as an old fridge, providing him a first comparison base between the live experiment and the simulation obtained from a derivative version of the model.

Doug then performed an experiment campaign. By comparing his simulations with the tests, he noticed that the fridge temperature was three times more sensitive in his model than in reality. This was very surprising, and this dramatically impacted on the consumption level accuracy.

After analyzing, he figured out one error during the mathematical modeling process. He wrote the wrong physical formula and introduced a factor of 3. After correcting, the results were much more satisfying, with ±5% accuracy compared with the experiments. This accuracy remained when studying different life situations.

Doug was confident in his model's validity level; he made a sequential analysis, then a satisfying general analysis. Then, he could confidently forecast the expected ±20% accuracy and thus make sure his model was valid in the domain under consideration.

4.3.7 Best Practice 7: Producing Useful Results

MISTAKE 7

We saw that Doug answered his client's project issue. Indeed, he predicted the new fridge's consumption with the expected precision. The project manager got back to him:

Project manager: "Doug, I'm proud of you. Your first conclusion was relevant. Now we would like to know the consumption of the old fridge in the same life situation. But this time we need more accuracy, at ±5%."

Doug: "With the current model, I can make this simulation, but not with such fine accuracy. If needed, I would have to build a new, sharper model, but it would take several months of work."

Project manager: "Unfortunately, we need the answer quickly. We will proceed with live experiments to get the answer."

We will see later that Doug actually had the opportunity to easily answer the actual needs of the project manager. He couldn't come up with the additional results that would have been useful.

Simulation teams often can't produce really useful results for their projects, or at least not as much as they could. Of course, the stakes in these matters are high. It is a shame to invest in simulation if, in the event, the actual outcomes are nonexistent or limited. Also, this topic is finally very complex. On the one hand, the need is sometimes wrongly expressed on the project promoter's side. On the other hand, the simulation teams may not always bear in mind the strengths and weaknesses of their own tools. Thus, some value creation opportunities can be missed.

How do we identify whole simulation studies the project could benefit and in this way produce the maximum simulation results?

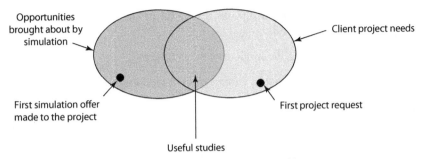

FIGURE 4.8
Managing to define the simulation studies as both achievable and useful to the project.

4.3.7.1 Being Aware of Project/Simulation Team Misunderstandings

On the project side, there is a set of needs. The project is scarcely aware of the opportunities offered by simulation. Thus, the first project demands are often studies that simulation can't easily cover.

On the simulation side, there is a wide range of opportunities. Also, simulation teams are often unaware of the project's actual needs (they only consider the requests, without the underlying needs). In this way, the first of the simulation team's propositions generally don't really match the project's needs.

The consequences of this misunderstanding are

- *Missed opportunities*: When project and simulation teams can't identify the studies that would have been relevant, or at least can't identify the exhaustivity of these studies.
- *Useless or even harmful studies*: A simulation study is sometimes initiated when it is useless. That situation comes from a mutual misunderstanding at the beginning of the study. In the best cases, they figure out the study is not relevant. In the worst ones, they don't notice that a simulation study isn't really useful (e.g., the precision doesn't meet the needs); the project may lead in the wrong direction.

This gap between the project vision and the simulation team is illustrated by Figure 4.8. Then the question becomes how to identify this set of useful studies in practice.

4.3.7.2 Suiting Project Needs and the Simulation Scope of Opportunities to Identify the Set of Relevant Studies

Knowing the project needs:

- This is about apprehending the project's real needs, which are often not stated first. We mentioned this before in Best Practice 1, and the communication aspects will be reviewed in the organizational best practices in Chapter 5.

Knowing and expressing the simulation opportunities:

- *Knowing the strengths and weaknesses of simulation*, as we covered in Chapter 3.
- *Knowing the validity domain of the model*, as we mentioned in the previous best practice.
- *Forecasting level*: Three levels can be reached.
  a. *Relative forecasting in a nonquantified way*: In this case, the model only manages to forecast in what way an output variable will change, depending on the model input variable's way of variation.
  b. *Relative forecasting in a quantified way*: In this case, the previous situation is reached, and moreover, the model manages to quantify the output variable change intensity, depending on the input variable change intensity.
  c. *Absolute forecasting*: In this case, the model manages to satisfactorily forecast a model's output variable value, according to the input variable value.

Suiting the result:

- A good understanding of the need and of the opportunities enables a better identification of the exhaustivity of the studies to start. For instance, a project often initially requests an absolute forecast result (level 3), whereas the model is relevant only as relative. In this case, a comprehensive work often allows one to notice that, in the end, the project has a need for a relative result and that defining the simulation objective in this way is appropriate.

Let's detail a few common traps:

- *Excess of confidence on the project side, wrongly held*: In this case, the project has an excessive trust in the simulation. This is usual in a complex model situation. This complexity can lead one to think that as the model is complex, it has to be pertinent. This may emphasize whether overfit correlations were made, which gives the illusion of accuracy. So, the validity model level must be openly shared.
- *Lack of confidence on the project side, wrongly held*: This is the opposite situation. In this case, the lack of confidence leads one to reduce the number of studies performed, and many opportunities are missed. As previously, spending time on giving a realistic vision of the model's validity is profitable.
- *Fool's game*: An excessive request for simulation accuracy may occur, accuracy that is in the end unnecessary to look after the project's

actual needs. A margin is kept because of a mistrust in simulation. On the other hand, simulation teams often have an overoptimistic message (deliberately or not) of their model's accuracy. In the end, the illusionary accuracy declared on one side suits the need for accuracy exaggerated on the other side, and studies are performed. This fool's game may work for some time but turns afterward, as soon as the balance changes into the impacts mentioned previously (pointless studies or even harmful, missed opportunities).

So, endeavoring to understand the project needs, as well as controlling the field of opportunities brought by simulation, enables the widest range of studies with useful results for the projects.

APPLICATION OF BEST PRACTICE 7

Doug became aware that he potentially missed an opportunity to add value and went back to the project.

Doug: "You asked me to forecast the last fridge draft consumption with a $\pm 5\%$ accuracy. Why did you ask for this value?"

Project manager: "We want to assess the consumption decrease, in a similar life situation, of the new fridge compared with the old one. So we requested a high accuracy of the old fridge consumption, as the new fridge value accuracy was only $\pm 20\%$. The accuracy of the delta, our greatest concern, will be low if both values are only $\pm 20\%$ accurate."

Doug: "In this case, I can provide you with the delta value right away. The model I built is very efficient in assessing a variation but much less so for getting absolute results."

Project manager: "Excellent! In this case, it is the delta value we want, with $\pm 20\%$ accuracy."

So, by understanding the actual need of his customer's project, Doug managed to define a study with relevant results, and this was without changing his model.

4.3.8 Best Practice 8: Maintaining and Storing the Models

MISTAKE 8

Thanks to the model he built, Doug managed to provide some simulation results that were relevant to the project. Three months went on without any new work requests for Doug. Nevertheless, during these 3 months, the new fridge conception evolved. Among other things, the project manager chose a supplier for the new compressor, as well as for the cooling loop controller. He then got back to Doug.

Project manager: "Doug, the compressor and the controller the supplier suggested to us are slightly different from those we told you about 3 months ago. Could you assess the electrical consumption gap related to these hypothetical differences?"

Doug: "Yes, but my model isn't that flexible, I'm going to need 3 more weeks to do that analysis."

Project manager: "Unfortunately, we need to make our final choice within 2 weeks, so we will run live experiments instead."

Time jump: Two years have passed. The "new" fridge is marketed, Doug got a new position, and Emily, a junior engineer, took his. A new generation of fridge is in development.

Project manager: "Emily, we are designing a new fridge. It will be even less energy consumptive than the current generation, with a more efficient compressor and isolation. Doug built a model 2 years ago, could you update and use it?"

Emily: "I found Doug's old data, but the model is incomplete and doesn't work. I will have to start all over again and build a new model."

So, by not properly managing the maintenance and storage of the models, Doug missed opportunities for value creation.

To avoid Doug's mistake, investment in model maintenance and storage must be considered. Maintaining the models is crucial, otherwise they become quickly outdated. Proper archiving is as important. Without it, not only does reusing the model later become impossible, but the traceability of the results that led to decisions isn't ensured. We are going to detail how to ensure good maintenance and good storage of the models.

4.3.8.1 *How to Handle Model Maintenance*

4.3.8.1.1 *What Factors Elicit the Need for Maintenance?*

- *System changes*: One of the most common reasons. In industrial companies, the definition of the studied system constantly changes. To depict it, these hypothetical changes must be considered when necessary.
- *Modeling changes*: After using a model for a while, some limits will be reported, and potential optimizations will be noticed and must be

included at the right time. These changes must be considered (non-essential to proceed with the study): continuous improvement, with the iterations covered in the previous best practices (these being critical to meet the study objectives).

- *Software or computer tool changes*: Software and computer tools often need to be updated. These updates may require a model change to keep it usable and compatible with the new versions.

4.3.8.1.2 *Identifying a Proper Maintenance Frequency*

- *Avoiding both extremes*: A lack of maintenance leads to the issues mentioned previously. On the other hand, maintaining the model too often is pointless. For instance, changing the model each time the system definition evolves is not only time-consuming but also probably useless. A project team may consider a solution A, then B, and then go back to A. If no study was asked for solution B, it would have been pointless to update the model in the meantime. In the same way, if the software is updated twice in a row, but a simulation isn't necessary meanwhile, changing the model after each update is irrelevant as well.

- *Finding the most useful maintenance frequencies*: To avoid both extremes, the project team must care about the actual needs for maintenance (by excluding useless maintenance and being aware of its importance). Then, they must assess maintenance cost levels (time, money, and means). By balancing needs and costs, maintenance obviously becomes rewarding.

4.3.8.1.3 *Anticipating maintenance*

- Anticipating the maintenance process is always good, right after the beginning of the model building. The model must be built in order to have it be easily flexible, changeable, and settable. A few working hours at the beginning will save many days of work later when it comes to updating the model.

4.3.8.2 How to Handle Model Storage

Knowing how to archive a model: What is archived is often unexploitable, even though the storage itself is done properly. How do we properly archive a model?

- The archiving necessarily includes
  - The system model (configured) and its environment
  - Everything required to run the model (including potential libraries)
  - The setting of a life situation to be experimented on (including the initial conditions)

- The expected results of this life situation
- A potential document resource (e.g., noting the model's accuracy level, the key hypothesis made, or even references to the possible correlations)
- The life situation to experiment with enables one to make sure the user obtains the expected results of the model.
- The following must be easily identified: the input, output, and internal variables, the system, the environment, the initial conditions, the parameters, and the life situation (otherwise, information must be included in the library).
- The archived file has to be self-supporting.
- Respecting configuration management is crucial (see the organizational best practices in Chapter 5).

Assessing the right storage frequency: As for maintenance, the right frequency must be determined. It takes time; it mustn't be done uselessly. Unfortunately, a lack of storage may lead to the same issues covered previously. A balance has to be found, which often consists in archiving the major versions (not necessarily all the transitional updates) or those dealing with key decisions.

Distinguishing model and result archives: These two are different, but the second one is connected to the first. There are several ways of storing data. The main issue is to archive both of them, avoiding redundancy and easing identification.

(Sometimes) following live experiment examples: Live experiment results are often better stored than simulation results. Two factors at least can explain this: live experiments have a deeper background and maturity, but they are also costlier, so more attention is paid to their data storage. Anyway, the methods linked to live experiments sometimes have to be imitated.

Storing for oneself as well as for others: The experience often reveals that opening an archived model after a year or more of storage is the same as seeing it for the first time. The archived best practices have to be followed, even if the user of the archive is also the author.

APPLICATION OF BEST PRACTICE 8

Let's get back to Doug before his position changed. Let's assume Doug was aware of the previous best practice.

The same 3 months passed without Doug receiving a request from the project manager. Nevertheless, this time, he anticipated the future request and inquired with the project manager regarding the upcoming deadlines and issues to come.

Once again, the project manager demanded an electrical consumption gap assessment with the new compressor and controller. This time, Doug was ready; he had performed a maintenance phase for his model in order to take into account the possible hypothesis changes related to the supplier suggestion. This way he could provide the project manager with results.

Then, 2 years later, Emily was asked again to start from the model Doug created. But this time, the model was properly stored, respecting the best practices, and Emily could use it.

Thus, by improving the model maintenance and storage, the simulation's value creation was maximized.

5

Efficient Use of Numerical Simulation: Organizational Aspects

Which organizational best practices should be applied to efficiently expand numerical simulation? Simulation working methods vary from one company to another, and they are often not standardized within the same company. Some methods are, however, more efficient than others, and some best practices are standard.

5.1 Stakeholders

Many stakeholders support numerical simulation expansion. Each of them plays a significant role, and according to how the work is managed, they become drivers or brakes. Let's start by listing them.

- *The simulation's final clients*: The top-level managers (or directors). We explained in Chapter 3 that the purpose of expanding simulation in industry is to encourage profitability. At least, head management has this concern.
- *The direct clients of simulation*: The project managers. The direct customers are those ensuring the systems development.
- *The model builders*: The simulation engineers. They develop the model.
- *The model users*: They use the model to create simulation results. Simulation engineers can handle this process, as can other interlocutors (including project team members or numerical simulation technicians).
- *The model creation team members*:
  - The top-level managers and project managers, who lead the teams
  - The numerical simulation pilots, who structure the simulation work in compliance with the other stakeholders

- The simulation experts (or simulation seniors), who can help and head the simulation engineers
- The numerical/IT support engineers, who assist simulation engineers with computer and numerical issues
- The project members or professionals of the field, who provide data or system information
- The suppliers, who bring technical support or models for specific systems
- The other modeling teams, if models are to be shared and integrated with other teams
- *Those impacted by the simulation*: Other players can be impacted by simulation without being involved in the simulation works.

We encourage stakeholders, if they are willing to help with the expansion of numerical simulation, to take note of all of this chapter's organizational best practices. Why?

- *To be able to train other stakeholders*: Numerical simulation is far from being perfectly developed and standardized within companies; several more years will still be necessary, as it keeps on renewing. Some best practices should be, for instance, the responsibility of project managers. If they know the best practices, they should be trained, or at least knowledge should be shared.
- *To be flexible and able to endorse another stakeholders' tasks*: Theoretically, each of them has a specific one. But in practice, depending on the company, the departments, and the projects, roles can vary. The simulation client can also build it, or the manager and the project manager can be the same person. This diversity comes especially from the low levels of standardization we mentioned previously. So, to be flexible, these best practices are a must.

5.2 Eight Organizational Best Practices

Eight organizational best practices can be listed. We have decided to highlight them with a case study. To this end, we will follow the experiences of Doug, a junior simulation engineer, and all the stakeholders he will have to interact with. They will make eight common mistakes and correct them using eight best practices, and so a short description of the usual mistakes will come before each best practice. Unlike in the previous chapter, the best practices aren't to be applied linearly, so we will take several steps back to reveal new issues.

CASE STUDY BACKGROUND

One of Doug's company's top-level managers became aware of the use of numerical simulation to design the next generation of fridges. That's why he took on Doug, who joined the design and application (D&A) team. Doug is the only numerical simulation engineer in the team. Simulation engineers work in another team in the company, sizing the subcomponents of the cooling loop.

5.2.1 Best Practice 1: Leading Change Related to Numerical Simulation

MISTAKE 1

So, Doug was hired, but for want of a vision, the company let him choose how to organize his work.

Doug figured out that the geometry of the fridge walls could be optimized by simulation, in order to decrease the fridge's heat losses and enhance energy consumption. He reported this to his supervisor, who agreed, although it wouldn't be applied in the middle term since it was already too late to review the fridge's geometry, which was currently in the design process.

Since he was the only one in command, Doug started the modeling work, acknowledging the previous chapter's technical best practices. Nevertheless, he quickly found himself isolated in his process. He needed his colleagues and the creators to take time and explain to him how the systems worked. But these designers were busy handling their daily emergencies. He could also take advice from simulation seniors. Unfortunately, they were related to other teams and didn't understanding the point of dedicating time to him.

Doug reported his issues to his manager, who advised him, but that didn't help him in the end. The manager was also very concerned by other issues and actually never believed that much in simulation. He only took Doug into his team because of top-level management pressure.

After having struggled for 4 months, Doug still managed to put together the fridge model. He showed that new geometries would decrease the fridge's energy consumption.

His results weren't used by the project managers for another year, in the fridge's next-generation design. Seeing the costs of a geometry change, the project managers decided not to use Doug's suggestion.

It has been a year and a half since Doug joined the company. The few matters he has handled have always taken the same path. When he requested more means from his manager, he got the usual answer: "It's been a year since we started expanding simulation, with time and money costs and no significant outcome: no decision has been based on simulation results. We can't spend any more energy on simulation."

So, the company couldn't make numerical changes or get any benefits through them.

Expanding numerical simulation implies a significant change for the company. We have already explained in the first few chapters how it induces a crucial disruption. This change needs to be handled, in order to benefit from its contribution. Even with this high potential and some great skills, very poor results can be brought about by this change being driven incorrectly.

How do we lead this change better? There are many books on this topic. Our goal isn't to substitute, it is to show how these techniques apply to numerical simulation–related change driving. We structure the process using the eight steps defined by J. Kotter in his book *Leading Change* (Harvard Business School Press, 1996):

1. *Installing an emergency atmosphere*: It may be difficult to rally all stakeholders at the beginning of the simulation expansion, as they are already struggling with many other matters. The natural inertia of men and companies tends to make the status quo last. Managing to install an emergency atmosphere, trying to play the emotional card, is highly mobilizing. In order to create this kind of feeling, the ideal is to mix the rational and irrational. For instance:

    a. A competitor is already using numerical simulation for a topic and has already moved forward.

    b. New impact projects pop up and their success relies on simulation.

    c. A change of regulations is coming, and they seem impossible to treat without simulation.

    Who is to initiate this feeling of emergency? There are several options. The impulse can come from a project manager, top-level management, a simulation engineer, or any previously mentioned stakeholder.

2. *Building a pilot team*: The strength of only one person, whoever he or she is, is hardly enough to satisfy an ambition. To expand simulation, building an efficient team of complementary members with a good reputation within the company is necessary. A proper team should involve at least

   a. A simulation engineer to perform the operational work (the simulation engineer is the only one involved full time).

   b. A project manager and/or director. The role first includes making sure all the members of the project and fields dedicate time to the simulation teams. This is a simulation sponsoring and credibility role. Secondly, the manager contributes to a vision of the topics to be treated thanks to simulation.

   c. A simulation expert (simulation senior) able to provide the simulation engineer with relevant advice but who will also help to build the vision of the topics to be treated with simulation.

3. *Defining a vision and objectives*: This is the role of the constituted team, to build a vision of the works to do. That approach must be both ambitious and realistic. We will get back to this point in more detail in Best Practice 2.

4. *Explaining the vision*: In order to dedicate and contribute stakeholders to simulation work, they need to understand the reasons and the framework defined in the previous step. Simulation can be seen negatively by those who don't understand it. Time must be taken to explain both points 1 and 3. The investment is often underestimated: team members contributing with motivation because they have understood the approach are much more efficient than others contributing because "they have to."

5. *Break down barriers*: Some barriers can obstruct simulation expansion either because of a lack of means or hierarchical blockages. We will get back to this point in Best Practice 4.

6. *Create short-term success*: Probably one of the most important points is managing to create the first short-term success (we'll see how to properly handle communication between stakeholders to succeed in Best Practice 3). This success is relevant for two reasons. On the one hand, they justify simulation expansion and become the fuel of the success to come. Good communication allows one to motivate the teams. On the other hand, success enables progressive learning: the simulation teams learn in the short term, with experience, what works and what doesn't.

A common mistake is to start with some ambitious simulation work straight away, without undertaking less important works beforehand. The teams may then lose breath, trust can decrease, and results may never be obtained. First, the investment would have been high, but no success would be achieved. Another obvious mistake is to undertake short-term outcome work but without succeeding, and multiplying this type of work without learning from past mistakes.

7. *Building on the first success*: The first results are essential to the value of numerical simulation. First results are meager, and multiplying the success is necessary to defeat the unavoidable inaction for any change. Once the first benefits of simulation surface in a project, new needs to be fitted emerge. One must adapt and use them to improve continuously.

8. *Anchor the accomplished changes*: The change has to become a standard. Numerical simulation then becomes the reference tool to deal with matters. This step is also broadly underestimated. Some interlocutors will change over time, new ones will appear, and the motivation will decrease as the emergency situations vanish. The previous success can easily be lost. The changes achieved must be entrenched by defining the fitting standards and procedures (as we will see in Best Practices 5 through 8).

APPLICATION OF BEST PRACTICE 1

Getting back to the story at the beginning of the study, we assume the stakeholders have acknowledged the preceding best practices.

For a start, the supervisor made sure Doug was surrounded by the right people. A simulation senior from another service was involved so he could provide Doug with advice. He and Doug decided to choose another topic. Indeed, the geometry change Doug considered wasn't an urgent or crucial matter. After inquiring about the context, they realized one of the competitors was designing a new generation of fridge with a variable-speed compressor, making it better at energy consumption. Doug's company was still using constant-speed compressors and decided to keep them like that because the prototypes wouldn't be available for one year. Doug found his subject: assessing the impacts of changing the compressor to a variable speed one, which may lead to a change of compressor reference, and this way catching up with the competition (case studied in Chapter 4). However, work needed to be started quickly.

Doug easily developed the model, following the senior's advice. Hesitations arose about the procedure, and the stakeholders took their time, but as they did, Doug took time to explain the relevance of it and sometimes used the senior's influence.

Doug found key results allowing him to validate the variable-speed compressor concept. The project manager decided to go for this solution and changed the compressor reference.

The project team became aware of the benefits of simulation, which soon became a must for conception, before the live prototypes. Many new topics were put forward for modeling.

Over time, new projects came up, and Doug was asked new questions. From then on, the team grew, and he was no longer the only one handling these simulations. To maintain the quality of work, he set up standards that all the numerical simulation engineers use now.

So, by leading the change efficiently, the company managed to dramatically increase simulation value creation, in both the short and long terms.

5.2.2 Best Practice 2: Defining a Numerical Simulation Expansion Strategy

MISTAKE 2

Let's go back a few months, when the first results brought about by simulation allowed one to value the project. This time, the company decided to introduce Doug to another department's simulation supervisor. Thanks to Doug's first achievements, this supervisor was highly requested. Within only a month, four new requests came up.

1. Assessing the cooling speed benefit brought about by improving the exchanger's efficiency
2. Assessing the electric consumption benefit brought about by changing the exchanger position inside the fridge
3. Comparing the energy consumption differences when using a constant-/variable-speed compressor on hundreds of life situations
4. Assessing the average duration of the variable-speed compressor for a standard customer

As the first request came 2 weeks before the others, the supervisor decided to deal with it first. Then, the other three came simultaneously. Unfortunately, the project managers were not the same. The one who made request 2 was particularly insistent, so the manager decided to prioritize it.

While he was right in the middle of Study 2, Doug realized it was very easy to perform a live experiment; this would have been cheaper and the result would have been more accurate. Did he have to inform the project manager? Furthermore, the work he did for Study 1 led him to develop an entirely new model, and weeks went on. Was this really the most urgent project? And in the end, were these four studies the only ones to consider?

So, for lack of a defined vision of the activities to perform, Doug's supervisor didn't best organize the modeling works.

When expanding numerical simulation, a strategy is required for two reasons.

First, a multitude of subjects can be treated by simulation, and choices have to be made: are they profitable or not? The choices have to be made voluntarily and rationally, at the risk of being randomly guided by the external demands that can come up and which appear to be irrelevant once performed (in which case, the credibility of the activity will be questioned). Second, it must be remembered that simulation is a tool, not an end. So, the reasons for simulation work to exist will often be questioned (see the previous best practice). The justification of the work must be strong to imply participation.

How to Build a Numerical Simulation Expansion Strategy

- *Choosing the scale of study*: The level of vision to be built must be established. Depending on the case, this can be performed at the scale of the company or reduced to the scale of a department.
- *Building a list of candidates*: This is about identifying all the subjects where simulation is relevant and so where it is profitable (see Chapter 3). Four methods can be used.
  1. *Exhaustive approach*: This first approach is more theoretical than practical but interesting if the scope is limited. It consists in balancing the need (present activities and forthcoming) and the offer (the opportunities offered by simulation) to assess its relevancy regarding some specific activities. Processing that identification requires one to gather both stakeholders who have a good knowledge of the company's activities and others familiar with simulation and its opportunities. As seen in Chapter 3, the desired benefits are
     - Reducing the number of live experiments
     - Improving the performance/product cost ratio
     - Stimulating and enabling breakthrough innovation
     - Reducing the design teams costs
     - Making the conception stronger
     - Decreasing the time to market

2. *Benchmark approach*: This time, the present and forthcoming activities must be identified again (segmenting it properly). Then, a comparison with the competitive companies has to be done to assess whether the company is falling behind or not on specific activities (keeping in mind that competitors may have made mistakes and what works for them may not work for the company). For instance, in the car industry, when a constructor erases a prototype phase, replacing it with more simulation, competitors often do the same shortly after.

3. *Time approach*: This one is often very effective but isn't exhaustive. It's based on the hypothesis that, at this time, numerical simulation has already been expanded within the company in every profitable field. In other words, if no change occurs, there is no new activity to expand simulation. Thus, there are new opportunities for simulation among the sources of change:

   – New company activities: for instance, in the car industry, autonomous car activities have generated a strong need for numerical simulation.

   – New opportunities brought about by simulation, to be covered in Chapter 6. Simulation techniques improve continuously. For instance, in the nuclear domain, computing power is allowed to go further in thermal-hydraulic modeling.

4. *Brainstorming approach*: One must obviously remain open-minded to any suggestion or proposition (or even provoke it), which can come from multiple sources. For example, some project managers might realize some of their activities could be covered by simulation. Also, noticing simulation expansion in another industry in a similar activity may encourage the consideration of new studies.

- *Prioritizing the elements of the list*: After identifying potential candidates, put them in order to assess which are priorities. One can't have everything. The criteria are usually

  - *Profitability*: How profitable would the simulation study be?
  - *Investment*: How much has to be invested (in time and/or money)?
  - *Success probability*: What are the chances of success?
  - *Deadlines (need and feasibility)*: Is the study to be performed in the short or middle term? What deadlines should be observed?
  - *Necessary skills/means*: Are the necessary skills/means available and at the correct rate?
  - *Synergy*: Are there synergies with some studies?

- *Building the expansion framework*: According to the needs and means available, after establishing a prioritized list of potential subjects, an expanding framework can be defined. This is to evolve over time, as new opportunities will come up.

The numerical simulation expansion strategy must be used to drive the work and to enable the maximization of simulation benefits.

APPLICATION OF BEST PRACTICE 2

Let's go back to the previous story, a few weeks earlier. Doug's supervisor just received four study requests. How could he build his expansion framework?

Simulation is new to fridge-designing activities, so the manager decided to build a forecast of present and forthcoming activities with the project managers and of simulation opportunities with the simulation senior.

He figured out there was currently only one fridge to be designed and that both the potential other needs were

1. A compliance assessment for the variable-speed compressor controller
2. An assessment of the gain in electrical consumption related to the condenser size change

So he decided to go through a six-alternative analysis using the criteria previously defined and obtained in Chart 5.1.

So, he decided to prioritize Study 5 and then Study 3. Then, regarding the deadlines, Studies 1 and 6 are prioritized. Finally, after a discussion with the project manager, he decided to exclude Study 2 because of its nonprofitability. Doug's manager could then set an agenda with the project manager.

Thanks to this expansion framework, the company managed to maximize the simulation-related value creation.

	Study number					
	1	2	3	4	5	6
Profitability	++	NOK	++	+++	+++	++
Investment	+	+	++	+	++	+
Success probability	+	+++	+++	+	+++	+
Deadlines (need and feasability)	++	+++	++	+	+++	++
Skills/required means	+	+++	++	+	++	++
Synergies	+	+	+++	+++	+++	+

CHART 5.1
Use of the reading grid to identify the priority level of each of the studies considered.

5.2.3 Best Practice 3: Managing Communication for Numerical Simulation

MISTAKE 3

Let's go back several months beforehand; the company had just hired Doug.

Once again, designing a new generation of fridge was considered, using a variable-speed compressor instead of a constant-speed one.

This time, the project manager wasn't the same, and he was a senior of the company. He decided to call for numerical simulation according to the director's suggestion and went to Doug.

Project manager: "Doug, could you build a model studying the feasibility of using a variable- instead of constant-speed compressor for our next generation of fridges?"

Doug: "Yes, but what studies will we need to perform with that model?"

Project manager: "There will be a multitude of studies to do; we do not have an exhaustive vision of it."

Doug: "But I need to know which ones in order to build a relevant model."

Project manager:

"I requested the preparation of a live prototype and I wasn't bored with these kinds of questions. I don't see why we should we have to give any further details in order to build a model."

Doug was a little upset by that comment. The project manager didn't get it; that was his problem. However, he began the modeling work.

After a few months, Doug completed an appropriate model. This took longer than planned as he had to restart several times. Indeed, it occurred that processing fridge electrical consumption assessment was expected, while the model Doug started wasn't adapted. Doug introduced the first simulation results as they were available.

Doug: "To sum up, in this life situation, the variable-speed compressor will reduce the electrical consumption by 30%."

Project manager: "To be honest, I hardly believe these results. Are they reliable?"

Doug: "Yes, regarding the compressor; I assumed all the fluid was in a dry steam state, and I used the fluid isentropic compression equations. Nevertheless, in order to take the internal friction into account, I included an isentropic coefficient parameter I identified in experiment results."

Project manager: "I'm not sure these technical terms mean anything to me, but what is the accuracy of your results?"

Doug: "Hard to tell, but I believe it is good."

Project manager: "I need a definite answer to decide whether to validate the concept or not; it will make us lose millions of dollars if it appears not to be accurate."

Doug: "Simulation isn't a tool that gives absolutely accurate results!"

In the end, the project manager decided not to take the simulation results into account. The project was aborted: waiting for the live prototypes was too risky in terms of costs and simulation didn't allow to remove that risk (when it could have).

So, because of a miscommunication regarding simulation, the company couldn't benefit from the gains it could have brought.

So, establishing good communication among stakeholders involved in any way in numerical simulation is crucial. Without it, even with skilled players and a true potential for simulation, success will not be achieved.

Efficient communication is often difficult to establish among these team members for three main reasons:

1. *Generation gap*
    a. *Seniors*: We saw in Chapter 2 that simulation is actually quite a new activity. Many senior company stakeholders, who are truly influential, were not trained during their studies. Although numerical simulation had already started expanding, it wasn't yet systematically taught. These stakeholders, mostly directors of project manager, are often suspicious of this novelty and wonder (legitimately) whether it is simply a valueless fashion or a deep and lasting change.

       Obviously, some of these people are more or less conservative.
    b. *Juniors*: Conversely, recent graduate engineers are almost always aware of simulation and its benefits. A large majority of them have already used it during their academic curriculum, and although they are not specialized, they are aware of its weaknesses and strengths and understand its functioning principles.
    c. *Confrontation or cooperation*: The dialogue between these generations may not be easy and may lead to confrontation if it isn't guided. Many seniors have a very pessimistic vision of simulation and expect to be convinced of its benefits.

On the contrary, juniors may feel frustrated by their elders' immobility.

2. *Virtual-based communication*: Almost instinctively, it may be difficult to believe in virtual results rather than in something seen in reality (e.g., live experiments with prototypes). This is a reasonable reaction, and the simulation engineer may question the validity of the model when he or she compares it with reality! Thus, it is often difficult to build trust toward these simulation results, or at least harder than with live experiment results.

3. *Simulation engineers' communication*: It is risky to make generalities, and we do not intend to stereotype. Nevertheless, if it is true that simulation engineers are mostly smart and creative, it is also true they are often less good speakers. This factor, combined with everything we covered previously, can make dialogue complicated for those who don't try to adapt.

We just covered the three reasons why communication can be convoluted among the stakeholders involved in numerical simulation. How do we handle these issues? This is what we are about to see, dealing with each of the three previous points.

1. *Connect generations*:
   a. *Seniors*: A raised awareness has to be in operation for the benefits of simulation to be understood (this is one of this book's topics). One of the potential means would be to make the pilot team (see Best Practice 1) take time to clarify the simulation's expanding framework and strategy. Convincing this generation ensures its sponsorship for the future, and teaches them to put things into perspective regarding this tool in order to have realistic expectations (i.e., neither too high nor too low).
   b. *Juniors*: They need to learn how to communicate about the strengths and weaknesses of simulation, saying what is and what isn't achievable. It is also important to get some distance and to understand the legitimacy of some blockages that may occur.
   c. *Strong cooperation*: Reestablishing communication between these two generations allows one to benefit from the strengths of each of them, and in this way, work more efficiently and lead the simulation expansion framework in the right direction.

2. *Making the virtual tangible*
   a. *For those who receive the results*: The previous points must be acknowledged in order to avoid underestimating the value of the results and to understand them properly.

b. *For those who produce the results*:

   i. *On the form*: A first obvious point consists in caring a lot about the form of the results to make them comprehensible and structured. Simulation often lacks professionalism, potentially because of its originality and the large proportion of juniors. The rough aspect of its results may strongly harm credibility. Also, all the necessary information must be provided to fully cover the context (information which is often systematically given for live experiments but hardly when it comes to simulation).

   ii. *On the substance*: First, to be credible, the matter of the system must be perfectly understood—for instance, by getting involved in live experiments. Also, the validity level of results must be mastered (see the technical best practices in Chapter 4) and a confidence index must be given and communicated. One must always be aware of the potential reading of results. Misunderstandings often occur and lead to wrong decisions.

3. *Communicating with simulation engineers*: Here again, stakeholders' awareness is crucial and will make teamwork efficient.

APPLICATION OF BEST PRACTICE 3

Let's get back to the moment Doug and the project manager met. This time, they were both aware of the previously mentioned best practices.

Project manager: "Doug, could you build a model in order to study the feasibility of using a variable- instead of constant-speed compressor for our next generation of fridges?"

Doug: "Yes, but in order to know where to direct the model's design, what studies will we have to perform?"

Project manager: "There will be a multitude of studies to do, we haven't got an exhaustive vision on them yet, but it will be focused on assessing the impact on the fridge's energy consumption."

Doug began the modeling work and focused his model on consumption issues. After a few weeks of work, Doug built an appropriate model and introduced the first simulation results as they were available.

Doug: "To sum up, the variable-speed compressor will reduce the energy consumption in this life situation by 30%."

Project manager: "I will need to validate the variable-speed compressor concept and several million dollars are at stake. From your point of view, what is your result accuracy?"

Doug: "This is hard to tell. The 30% accuracy I announced is around
±5%; I could make several validations with former fridge
versions. What is absolutely certain is that the benefit will be
at least 20% and at best 40%."

Project manager: "Fair enough. A 15% minimum gain is sufficient to
validate the concept. So we will keep the project going."

So, by establishing proper communication regarding numerical simulation, the company managed to combine everyone's strengths and,
in this way, benefit from the simulation's contribution.

5.2.4 Best Practice 4: Provide Any Necessary Means to Numerical Simulation

MISTAKE 4

Let's go back a few weeks earlier, when Doug began designing the
fridge model with a variable-speed compressor.

To set his model, he needed to perform a few live experiments with
the former generation of fridges (see the technical best practices in
Chapter 4), but the answer he got was irrevocable.

Project manager: "The D&A teams are currently overloaded and do not
have time to perform live experiments; that will generate
additional expenses. Also, we decided to expand numerical
simulation to decrease the number of experiments, not the
other way around!"

Doug didn't manage to convince the project manager to prioritize the
experiments he requested. He finally decided to use live experiments
that were performed previously. First, it took him a few more weeks,
and second, the accuracy he got for his configuration was lower and *a
priori* not sufficient to answer the study properly.

Then, he noticed his model was too heavy for the computer he had at
his disposal. The memory was too small, which made most of the simulations he ran fail without any results. Again, the answer was irrevocable.

Project manager: "We do not have any budget for new computer tools; you
will have to do without, whatever the consequences may be."

Doug kept working with his current tools. In the end, his results came out way too late for the project, with insufficient accuracy to treat the study, so much so that his work turned out to be useless.

So, by not mobilizing all the necessary means, the company failed to benefit from the contribution brought by simulation.

Making available any necessary means and taking down any organization-related potential barriers are crucial for simulation expansion. This is even more true at the beginning of its use, when it is still fragile and may vanish, even though it has achieved a few successes. Companies neglecting this factor, wanting to make small cost cuts, scuttle an activity that could have brought significant outcomes.

The four main means that need to be put at disposal are the following:

- *Staff*: Enough staff must be mobilized.
  - *Availability of the key stakeholders*: As we mentioned before, a multitude of stakeholders are involved in simulation. Some of them need to take some of their time to the simulation teams. For instance, simulation engineers will need to interact with the D&A teams, in order to properly understand the systems and model them. The nonavailability of the stakeholders may be involuntary (they have other emergencies to deal with) or voluntary (a powerful player is at stake, and they want to see the process fail). This matter must be dealt with at the very beginning of the simulation work and some of the key stakeholders' time must be made available to the simulation teams.
  - *Availability of the simulation engineers*: Also, simulation teams must be properly sized, especially in growth periods.
- *Experiments*: Simulation teams almost systematically need specific experiments for their work (see the technical best practices in Chapter 4). This need is frequently perceived in a negative way as simulation is precisely supposed to make the number of experiments decrease. These expenses must be considered as investments allowing one to decrease the number of experiments in the middle term. So, it is often necessary to spend money on these experiments but also on potential technical support for the organization (this relates back to the previous point).
- *Softwares*: The use of some specific software program is now unavoidable for a major part of simulation. It will also induce costs, which can be neglected, especially if the activity grows. Also, according to the type and number of licenses the company acquire, it can be more or less available, which can dramatically break the work.

- *Numerical tools*: Having numerical tools is obviously necessary to use the models built. These tools can go from simple personal computers to expensive, powerful remote computing servers. Providing the necessary means is crucial, otherwise the studies can be slowed down or the models made unusable.

Thus, it is crucial to make any necessary means available in order not to scuttle the efforts. Of course, the means we have just mentioned need to be rationalized. For instance, some software types can be requested that are useless in the end. Long lists of experiments can be generated. A wise arbitration will maximize the final profitability of the simulation activity. Finally, the stakeholders who provide these means are very diverse, and they rely on the precise structure of the activities. Let's mention that they generally should belong to the pilot team we defined in Best Practice 1. Thus, they can have the role of, at least, a sponsor to promote the provision of these means.

APPLICATION OF BEST PRACTICE 4

Let's get back to the previous situation, when Doug was requesting additional live experiments.

Project manager: "As the experiment teams are currently overloaded, it will be complicated to make new experiments, and this will cost money. I'll send you the experiments schedule; perhaps will you be able to include the ones you need."

After referring to the schedule, Doug identified an experiment similar to what he needed. It was certainly not perfect and it would generate a bit of extra work, but it wasn't enough to question the experiment schedule. So he managed to do his configuration.

Then, again, he figured out that new numerical tools were required to perform these studies and got the following answer:

Project manager: "We didn't anticipate such expenses for new computing tools, and we need it to be vital to grant it."

Doug: "I can do without it, but this will induce a 10-day delay, and I'll have to spend twice as much time on the study. Also, the tools we already have are undersized; it is a long-term investment to get new tools now."

Project manager: "Regarding the deadlines and the long-term outcomes, we are going to purchase these performance tools."

So, by mobilizing the appropriate means, the company generated simulation works that dealt with the problem.

5.2.5 Best Practice 5: Industrializing Numerical Simulation

MISTAKE 5

A few months had passed since the project manager received Doug's first simulation results, and the project stakeholders already understood the potential of simulation.

Project manager: "Doug, we are initiating a new phase; we're going to need many more results with your model. We will need the following results:

- For the 3rd of May: fridge's electric consumption results for a certification range (a specific life situation)
- For the 15th of May: fridge's electric consumption results for a standard client range (a set of 30 common life situations for the fridge's regular customers)

Now we have a fridge prototype, your model has to use the same hypotheses as the prototype."

Doug began by referring to the team who built the prototype and asked them to share their hypotheses, so he could include them in his model. He was told he would get them on the 20th of April. This didn't give him much time, but it was enough. In the meantime, he got the certification life situation coding ready.

Nevertheless, on the 20th of April, no hypotheses came.

Prototype supervisor: "We settled the hypothesis at the very last moment and it can still change the day before our experiments! You will get the set hypothesis on 1 May."

This was an issue for Doug; he knew he couldn't come up with the results in time for the 3rd of May. Nevertheless, he had already encoded the certification life situation; he wasn't going to give up at this point. He managed to get the results with a few days delay.

Project manager: "Thank you, but we realized that we actually didn't need these simulations, since experiments were performed on the certification range with the prototype on the 2nd of May."

Doug understood all the efforts he made were for nothing, but he didn't lose sight of the 15th of May deadline. The team then provided him with the standard client range. He incorporated the 30 life situations and got the expected results the day before the deadline. He noticed the consumer team's range didn't fit the standard customer's, but highly demanding clients.

Consumer team supervisor: "Indeed, there was a mistake in the data we provided you; we'll send you the appropriate standard client range again."

Doug's time was wasted. A few days later, the project manager got back to him.

Project manager: "We really need your results, even if it is incomplete, because we need to make a decision within 2 days."

Doug doesn't like to leave a job half done. In only 3 days, investing himself entirely, he managed to complete the study and delivered it to the project manager with a little delay.

Project manager: "Unfortunately, these results are no longer useful to us. We had to make a decision yesterday, so we performed live experiments with the prototype."

So, the company didn't manage to industrialize numerical simulation and missed several value creation opportunities.

As soon as numerical simulation has won its spurs with a new issue, industrializing it in order to use its full potential is important. Indeed, it implies a wide range of interactions among several interlocutors and activities, and the time comes when optimizing these interactions is crucial, in order to boost productivity.

How do we make an activity as efficient as possible? It is often not about adding extra work but, on the contrary, cutting it down, going toward more simplicity, and only keeping the works that really contribute to the value creation. There are already many books about that, written by Taylor, Ford, Toyota, and so on, not to mention the numerous manuals that exist on *lean*. Our goal isn't to supplant all these references but to show how these techniques can apply to numerical simulation. These are often commonsense practices. However, they are hardly ever applied. Why? Two factors could explain it.

First, numerical simulation is quite a new activity, so it isn't the first one to be considered when it comes to avoiding waste (despite strong potential). Secondly, in terms of affinity, numerical simulation tends to attract technical profiles inclined more for R&D than for efficiency, rationalization, and usefulness research.

We will see that in order to industrialize numerical simulation, relevant work processes must be defined and kept and applied efficiently.

How do we define and keep these relevant work processes? There are three steps to follow:

1. *Defining the current process*: By taking a snapshot of the current situation that will include the interactions among all the stakeholders defined in Section 5.1.

    To define that process, the key questions to ask are the following:

    a. What are the input data? For example, information, experiments results, subsystems models.

    b. What are the output data? For example, specific life situation simulation results, models.

    c. What is the added value?

    d. What are the means required?

    e. What is the monitoring system?

    f. Who is monitoring it?

    g. This snapshot often enables us to detect some shortcomings and is a first step for the next point.

2. *Improving the current process*: Several best practices are valuable.

    a. *Ensuring the process is consistent with those of the other stakeholders*: For instance, the simulation teams often expect input data at a specific date that isn't set on time, as the process of the team supposed to provide it has another schedule or, even worse, hasn't planned to provide it at all. These situations do not come from a wrong application of the process but from a matter of inconsistency. As soon as these issues become complex, or constantly changing, these situations occur much more that one can think.

    b. *Avoiding duplications*: Although the duplication can be more or less marked, the issue is very simple: the same activity is performed twice at least. Some duplicates may exist without being noticed by the stakeholders (then the same model can be built twice in two different places or at two different times). Sometimes, the stakeholders are aware of it but nothing changes, for good reasons or not. However, these duplications must be avoided and the process must be rationalized. In the 3D field, for example, meshes need to be created. In the past, for each type of study (mechanics, thermodynamics, etc.), due to the specific constraints, a specific meshing was created. Also, each team used to build their own models to meet their needs. Nowadays, models are often shared, built in the most flexible way to avoid building a new one that in the end would have been very close to the others.

3. *Keeping the process relevant*:

   a. *Freezing the optimized process*: The processes that have just been optimized need to be made official and written down. Otherwise, good progresses are very commonly made without freezing and are lost afterward.

   b. *Constantly upgrading the process*: The process-improving step mentioned previously mustn't be performed only once. The context changes and so do the activities, and the process that was appropriate the day before may not be so the day after. A proper pace for process upgrading must be found in order to keep it relevant, without changing it entirely every time a new topic comes up.

We have just covered how to define and keep the numerical simulation process relevant. How do we apply these processes on an everyday basis? Here are three best practices:

1. *Identifying numerical simulation pilots*: Historically, numerical simulation was an activity requiring very scientific profiles. Today, companies that succeed in industrializing numerical simulation have figured out the necessity of involving project supervision profiles to apply the process described above. For instance, to have the input data released at the right time, handle unavailable resource issues, and negotiate with the projects for the output data deadlines, a project supervision profile is required. This will be explained in Best Practice 6.

2. *Scheduling the process*: The numerical simulation pilot needs to plan work, according to the new topics coming up, using the definition of the generic process we just covered. The simulation activities schedule is often lacking. However, many scheduling best practices, which are mature in other fields such as production, all make sense here. A realistic and relevant schedule must be set that is convenient for the different stakeholders and prevents overloaded or underloaded periods for the teams, bringing results whenever they are useful. This schedule must then be monitored and adapted in case of potential drifting.

3. *Monitoring the work by avoiding waste*: We will mention here a few common mistakes to prevent that limit the added value of simulation.

   a. *Producing unused models and results*: Simulation teams often produce models and results that end up not being used in the end. Several scenarios may occur. First, a project may request numerical simulation, and if the need is incorrectly defined or miscommunicated (see Best Practices 1 and 3), this may lead to useless work. The numerical simulation pilot is meant to prevent this. This can also come from an incorrect definition of the appliance of the schedule. Then the results come too late. This is common, as simulation teams often enjoy building models, while their utility comes second. They focus on the model's development quality rather than on deadline observation and the impact on the projects. On the other hand, project teams often misunderstand the simulation activities, which induces a false vision of the progress made (see Best Practice 3), and they fail to prevent this drifting, so much so that if the numerical simulation driver can't handle the situation (or if there isn't one!), the simulation doesn't make any contribution.

b. *Changing hypotheses too often*: As a project goes on, some hypotheses change (including the system's size, its use situations, etc.). Simulation teams are then asked to take these changes into account; otherwise, the simulation results won't include updated data. If they try to respond to all these requests, they can end up overwhelmed by an excess of changes, so that the final results never come (or come after the deadline), which is worse than having results fitting slightly updated hypotheses. Filtering these project requests is crucial (once again, this is the numerical simulation pilot's duty). A compromise needs to be found between the project's need and what can be achieved through simulation. In order to convince oneself that simulations do not need to be constantly updated, all that is needed is an analogy with live experimental results. The prototypes that are built, because of the time they take, are never updated regarding the project hypotheses. One must be aware of the gap between the project vision and what is actually included in the model in order to see the results with perspective.

c. *Using noncompliant input data*: Another very common drift is the reception of noncompliant input data by simulation teams. In the best case, the team figures it out. A first integration work has usually already been made, and the wasted time corresponds to the time required to figure it out. The noncompliance of the input data can be multiple: a submodel (which is input data) isn't delivered in the correct version or some libraries are unavailable or the experiment results (which are input data as well) do not fit the configuration expected. On the contrary, in the most challenging cases, the team doesn't realize the input data are noncompliant and time isn't the only thing wasted. Indeed, in this case, the mistake can impact the output data and lead the project to wrong decisions. It is crucial for the input data to be correct in the first place. In order to make sure of that, a suitable monitoring system must be set up and team members must be held responsible for the quality of their output data (which are the input data of others). For instance, before sharing a model, its compliance has to be checked through a test case (this point will be covered in Best Practice 7).

So, managing to have a relevant numerical simulation process, and making sure of the resources to have its application mastered, enables one to industrialize numerical simulation and shows its full potential.

APPLICATION OF BEST PRACTICE 5

Let's go back to the previous situation, after the project manager had defined his expectations. This time, the company decided to choose a numerical simulation pilot (who acknowledged the previous best practices), whose only duty was to coordinate the team activities.

The pilot created a very simple schedule after referring to all the stakeholders, including the prototype team. He got back to the project manager.

Simulation pilot: "As it is, we won't be able to get the results on the certification range for the 3rd of May. Indeed, it doesn't match the prototype team schedule. Can you delay the deadline by a few days? Otherwise, we can try to figure out solutions that will reduce the schedule, but they will potentially require additional means."

Project manager: "Actually, the request for the simulation results in terms of certification range isn't too important since we already have the prototype results. So we can cancel our request."

This way, the simulation team avoided a useless study.

Then came the next study. The simulation manager took the care to warn all the stakeholders to provide the simulation team with accurate input data and prevent useless work and two-way trips. He also asked Doug to quickly check the compliance of the input data he was sent before using them, especially those with a major impact on his study.

Doug then figured out, right after he had received the standard customer range, that it actually matched the very demanding customer range. The simulation supervisor referred to the consumer team to gather more suitable data and a stronger means of data exchange, preventing this kind of mistake from occurring again.

So, Doug was able to come up with the results on time for the project and thus prevent an expensive prototype experiment phase. Numerical simulation entered a manufacturing process regarding this matter and was able to keep on expanding, offering full potential.

5.2.6 Best Practice 6: Managing the Numerical Simulation–Related Skills

MISTAKE 6

Several years went on. Numerical simulation kept on expanding over new topics, and a dedicated team was created, including 10 numerical simulation engineers.

A fridge project manager went to see the modeling and simulation (M&S) team supervisor.

Project manager: "I need to assess the average annual electric consumption of our fridges. We just went through a customer survey, and 100 life situations were defined. We would need you to forecast the electric consumption in each of these life situations, in order to assess an average value."

Simulation manager: "Our models will allow us to perform such a study. But this work will be very repetitive, since it will consist in repeating basic work 100 times. I'm in a dead-end situation. My simulation engineers do not accept having to perform this kind of boring study anymore; another one quit last month."

Project manager: "You don't know how to industrialize this kind of simulation to make the work less repetitive?"

Simulation manager: "This requires experience. Doug would have known how to do that. Unfortunately, he quit the simulation team a year ago. He was considering a numerical simulation senior career, but since this position didn't really exist within our company and his salary wasn't increasing, he ended up quitting."

Thus, because of poor numerical simulation–related skills management, the company couldn't benefit from its contribution.

To get the most value from numerical simulation, the right skills need to be hired and efficient work needs to be enabled. How? This is the following topic.

In the first place, one of the prerequisites is to understand that there are multiple kinds of skills that are necessary to succeed in expanding numerical simulation. The main ones are

1. *Numerical simulation engineer*: Designs and builds models, uses them to produce results.
2. *Numerical simulation pilot*: Plans simulation-related work in coordination with the other stakeholders.
3. *Numerical simulation senior*: Advises the simulation engineers, to make them benefit from his experience.
4. *Numerical/IT support engineer*: Assists the simulation engineer in numerical and IT issues.
5. *Numerical simulation technician*: Uses the models to produce results. In some cases, model-based results production work can become very "in the loop," even repetitive, to the extent that there is no need to mobilize the simulation engineer any longer. Unfortunately, this work can become close to assembly line work on a computer. This kind of work needs to be as reduced as possible.

We just went through the five kinds of skills that are most required. Five skills doesn't mean five people, as a same person often fulfills several of them. Also, we only mentioned here the skills directly related to numerical simulation, without including either the manager (who can be the simulation manager or the expert) or the potential live experiment managers (who may be involved in simulation or the other way around).

Now we have defined the necessary kinds of skills required, how do we recruit the right persons and provide them with the proper means to work efficiently? This is the upcoming topic.

5.2.6.1 How to Recruit the Right People

The right profiles must be identified that fit the following kinds of skills:

- *Numerical simulation engineer*: The first and main wanted skill is the ability to use the scientific approach with structure and rigor. Then, some other important points are much easier to develop over time: a knowledge of physics related to the studied systems and knowledge of the simulation software used. Curiosity and creativity are also both valuable assets.
- *Numerical simulation pilot*: Here, the most important thing is to seek out pilots who are able to coordinate both activities and stakeholders. So, the position requires human qualities as well as the care to build a relevant organization.

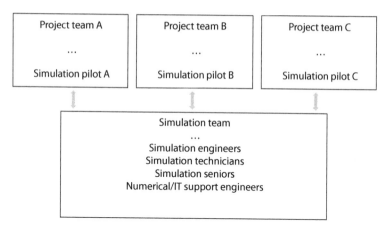

FIGURE 5.1
Example of numerical simulation stakeholder's team allocation.

- *Numerical simulation expert*: The expected skills are the same as those of the simulation engineer, but with many more years of experience, enabling him or her to have perspective in any occasion.
- *Numerical/IT support engineer*: The main expected skill is a mastery of the numerical and IT tools in question. Also, good interpersonal skills will guarantee oral communication efficiency.
- *Numerical simulation technician*: The demands in terms of technical skills are minor. Mainly rigor will ensure the robustness of the results produced.

5.2.6.2 How to Allocate and Gather Those Skills

- *Allocating the skills*: At the early expansion of numerical simulation, the five previously mentioned skills often happen to be gathered in one person, so skills allocation doesn't really matter. The person in question often joins the project team. When activity increases, it is time to think about which teams to allocate persons/skills to.
- *Connecting simulation skills to the project or not*: When several company projects use simulation as a tool, the idea arises of gathering the different simulation skills for each project or connecting them to each other (these are then separated from the projects). The opportunities are wide, and we depict an example in Figure 5.1.
  - *Advantages/disadvantages*: There are advantages/disadvantages to connecting the project simulation skills. Being closer to the projects (e.g., being a member of their teams) enables one to have proper knowledge of the studied systems and then build relevant models, close to reality. On the other hand, tying in with simulation skills

allows one to better pool the works and strengthen simulation skills, which will profit the projects. The best compromise has to be found.

- *Project closeness hierarchy*: Some simulation stakeholders will naturally have to be closer to the projects than others. For instance, the simulation pilot must be clearly aware of project issues and is often the first one to join project teams. On the other hand, there is no use connecting numerical/IT support engineers to the projects. In general, the priority of closeness to the project is in the following order:
  - Numerical simulation pilot
  - Numerical simulation engineer
  - Numerical simulation technician
  - Numerical simulation senior
  - Numerical/IT support engineer

5.2.6.3 *How to Retain and Convey These Skills*

- *Considering numerical simulation as a major domain of expertise*: Many companies have developed expertise in domains (e.g., mechanics or electronics) that are significant for their activities, but this is not yet the case in numerical simulation. However, numerical simulation is a full-fledged technical domain, with its own techniques and knowledge. This is also a serious investment for many industrial companies (see Chapter 3) and deep mutations are still in progress (see Chapter 6). Companies need to create (if they have not done beforehand) and maintain this numerical simulation expertise domain (with all that this implies: nominating baseline speakers, providing access to training, communicating with universities and other companies, etc.).

- *Expanding and maintaining this expertise*: One of the major stakes for the true development of numerical simulation expertise is to enable engineers real career prospects. Indeed, nowadays, numerical simulation is often just a transition for many engineers. Lots of them are junior engineers who were hired right after the end of their studies. They resign from numerical simulation after a few years. Training true seniors who will emanate the right aura is very difficult. Why do engineers resign from numerical simulation after a few years? Working as a simulation engineer is often underestimated and is not considered an IT management position (crucial decisions are made though). Simulation engineers' salaries are often lower, so is their acknowledgment compared with similar positions (inside or outside the company), and they end up shifting to new jobs. Companies succeeding in expanding and maintaining this expertise will have provided their engineers proper (rewarding) careers in numerical simulation.

APPLICATION OF BEST PRACTICE 6

Let's go back a few years, before Doug left the simulation team. At this time, the company was aware of the importance of initiating numerical simulation as a domain of expertise.

So, Doug was offered a job with in-house training programs to become a senior simulation engineer. He consequently decided to stay on the team, to keep on gaining experience.

Once again, the project manager asked the simulation pilot for a study to assess a fridge's average electric consumption in over 100 life situations.

Simulation supervisor: "We can perform that study. I will delegate it to one of our simulation engineers and check the validity of the results with a simulation engineer. Also, if this kind of study comes up again, we will industrialize the work, thanks to Doug and our IT support engineers." Thanks to appropriate skills management, the company benefited from the contributions of numerical simulation.

5.2.7 Best Practice 7: Managing the Models Expansion

MISTAKE 7

A few months passed, activities developed, and numerical simulation kept on expanding. From then on, four simulation teams were joining forces for the design of the fridges.

- A team modeling the fridge internal thermics
- A team modeling the cooling loop
- A team modeling the fridge command control
- A team assembling and synthesizing these three models

The synthesis work was led by the previous simulation pilot, who was unfortunately facing issues about to recur that he couldn't handle.

- *The team modeling the internal thermics*: The models were always delivered on schedule, but they often didn't work and couldn't be run by the synthesis team; they even showed different results. The simulation engineers often had to take time to figure this out and they also had a hard time detecting it systematically.

- *The team modeling the cooling loop*: The team stated the loop system was highly confidential, so they encrypted the model into a *black box* before sending it. The encryption took them some time, leading to a few unavoidable delays. Also, as soon as numerical issues arose with that model, the synthesis team was tied in with the other team because no change was conceivable.
- *The team modeling the command control*: That team was external to the company as they were working for the supplier of monitor controls. They were continuously changing their model interfaces, so much so that the synthesis modeling team spent a lot on new deliveries to adjust the new model and integrate it.

All these models' distribution and exchange problems curtailed the expansion of numerical simulation within the company.

Industrial companies need proper management of the distribution and exchange of models to expand the efficiency of numerical simulation. In the opposite case, that expansion may be blocked, or at least the benefits from simulation may be cut. How do we properly handle the models' distribution? This is the following topic.

One of the prerequisites is to acknowledge with whom the models need to be exchanged. There are five common cases:

1. *Exchanges with the project teams*: The purpose of a model built by a simulation team is sometimes for it to be used directly by the project stakeholders (the model simulation results are produced by the project stakeholders).

2. *Exchanges with other simulation teams*: This happens when the simulation teams integrate submodels built by other simulation teams. This is becoming more and more common (see Chapter 6).

3. *Exchanges with the suppliers*: This case is close to the previous one, but the submodel comes from an external company, the supplier. This submodel is the model of the system developed by the supplier. This situation is also more and more common (see Chapter 6).

4. *Exchanges with the customers*: This is the symmetrical situation to the previous case.

5. *Exchanges with the subcontractors*: Designing a model can be subcontracted (commonly by a modeling company). The model is sent to the company once its design is completed.

Now that we have covered the potential exchange channels, we can mention the usual issues and their solutions. These issues can be technical or nontechnical.

5.2.7.1 Technical Issues

- *Incomplete sharing*: This is by far the most simple problem. For instance, the shared model doesn't include any configurations or libraries. To prevent these mistakes, the teams must follow the same best practices as model storage issues (see Best Practice 8).
- *IT incompatibility*: A model often works on one computer and not on another one. Some software versions can be the cause as well as hardware configurations. The reader will refer to Best Practice 5 to solve these problems technically. Anticipating these inconveniences allows one to ensure the model exchanges work at the first attempt.
- *Lack of test cases*: A multitude of factors may result in a model functioning failure at the receiver's end despite it working at the sender's. The model can even show different results. We just explained that numerical/computer issues are one of the reasons. As soon as the model becomes complex, teams can take a lot of time to notice gaps between the sender's results and the receiver's. One of the most efficient ways to prevent this is to use a test case. How does it work? The test case includes two things: first, a life situation to test, as well as a model configuration; second, the model results using this configuration obtained by the sender. So, the model receiver can use the test case to make sure the obtained results are those expected. Two observations can be made. First, the life situation must be chosen wisely. It must be short enough to be performed quickly but long (or complex) enough to include different system stresses (and then the test will be comprehensive enough). Also, the life situation needs to be as close as possible to the actual life situation in which the model will be used by the receiver (not by the sender), to ensure as early as possible the reliability and robustness of the model. A second possible observation is that the test case isn't sufficient to exhaustively validate that the model will behave the same way at the receiver's end. In practice, this kind of issue hardly ever occurs if the case test was passed by the receiver. So, setting up a test case system is very efficient when guaranteeing the validity of the exchanges of models.
- *Lack of robustness*: This time, the model exchanged seems first to function satisfactorily but appears to be very sensitive to a range of factors (e.g., the life situation simulated) and fails if these factors are changed. The model becomes unusable in practice as it only works on specific points, which is insufficient for the work. The cause of the problem can be strictly technical, and it is technically difficult to make a robust model in different use cases considering the phenomena at

stake (again, see Best Practice 5). The origin of the issue can also be a lack of communication at the beginning of the study: the person who built the model wasn't aware of how the model would be used. This matter needs to be properly discussed at the beginning of the study (we are going to explain this point below).

- *Lack of standardization*: The model exchanges can become dysfunctional because of a lack of standardization. Two cases often occur.

  - *Model input/output data*: In the first place, the model's input/ output data can differ from those expected. This often happens when the conditions are not properly pointed out during the first exchanges. For instance, the input data can be the system speed and the measurement in km/h. Still, the speed can be expected in m/s. In the best case, it is noticed and just requires a small adjustment. In the worst case, it is not noticed, leading to incorrect results and poor decisions. There may be plenty of variations. The model input/output data are often not those expected if this is not observed at the beginning of the study. This is an issue when industrializing numerical simulation. To prevent this kind of problem, the expected interfaces have to be defined at the beginning of the study (i.e., name, unit type, description). This upstream extra work will avoid many mistakes afterward.

  - *Rules of simulation*: Another reoccurring issue is the gap between the rules of simulation (or the absence of rules) used by the model sender and the model receiver. Most simulation teams end up defining a set number of rules of simulation in order to work efficiently. For instance, if different solvers are available in a software program, a preferential solver is chosen (as one model developed with a specific solver may not work with another). These rules must be carefully chosen among the exchanging parties, or at least the gap between those rules must be tracked to reduce the necessary adjustments.

5.2.7.2 Nontechnical Issues

This time, the model exchange–related difficulties do not come from a technical issue. The potential will be described as follows:

5.2.7.2.1 Lack of communication

As we mentioned before, this may lead to problems during model exchanges. We have already detailed in Best Practice 3 the communication aspects linked with simulation. We focus here on its impact on model exchanges.

- *Upstream*: We saw that at the beginning of the study, teams often miscommunicate their expectations regarding the model. In this case, the obtained model doesn't match the one expected. Although this may seem tedious, properly defining one's needs at the beginning of the study, as well as the software, the computer, and the life situations, is crucial to prevent any surprises.

- *At the exchange of the model*: The model sender often delivers it as a consumer good, without any further explanation. A model is a complex tool, able to produce random results if misused. Model delivery must be paired with training (even quickly) to ensure the receiver will be able to use it.

5.2.7.2.2 Privacy Problems

- *Source of the problem*: This is one of the most difficult problems to solve concerning the exchanges of the models. A model is a representation of a system usually resulting from a very fine understanding of the system, with a large amount of information considered sensitive that must remain private.

- *Problems with a supplier*: The most obvious case is a model exchange with a supplier, who often shares a model of his own system (if it is paid for). Still, if the supplier agrees to share information concerning the system behavior (i.e., how it will react in specific situations), he won't necessarily share how that behavior was obtained. For instance, a supplier producing aircraft engines will probably agree to describe the engine behavior for any situation but will be much less likely to give the aircraft constructor all the engine features in action to produce that result (yet all these technologies at stake are often depicted in the model and so are visible). So, how do we share a model and maintain privacy at the same time? The most common solution is the black box. The model is encrypted, becoming nothing but an executable. The model reaction can no longer be understood, and privacy is maintained. Still, these black boxes can often cause many issues. First, ensuring the validity of the model is often more complex as it is impossible to explore it. For instance, the input/output definition problems mentioned previously can arise (checking an input is the right one is trickier if one can't explore its use within the model). Communication has to be perfectly managed (which is complicated to set). Black boxes also cause IT problems: integrating them into the model is complex and usable solvers are limited. So if black boxes offer a solution for privacy problems, it isn't optimal. There are three options when black boxes are not adapted. The first one consists in negotiating to get an open, unencrypted model (some suppliers accept this whenever the model doesn't include sensitive information). The second would be to ask the supplier for an open but simplified model (i.e., without any sensitive information). The remaining inconvenience is the simplification level and the extra costs to design such a model. The last solution is asking for a partly encrypted model, where only the encrypted areas contain confidential information. One of these solutions will help to solve the privacy problem with the supplier (these problems/ solutions also apply to a customer/company situation).

- *Problems with a subcontractor*: These problems can also occur with other stakeholders. For instance, when a subcontractor intends to build a model, he will need to thoroughly understand the system, including, once again, sensitive information from the company. Then, it must be decided whether to unveil confidential information or to make the model internally.

- *Internal problems*: Some information may be confidential even inside the company itself. Privacy can be managed even for employees to strengthen protection barriers. What was mentioned before with the supplier, client, or subcontractor applies again by analogy to internal suppliers, clients, and subcontractors.

5.2.7.2.3 State-of-Mind Issues

- *The model expansion is stalled*: Some simulation teams can be really unwilling to share their models internally, for good and bad reasons:
  - *A good reason*: One of the obstacles to sharing models is their incorrect use. Producing incorrect results is very easy, even with a relevant model. We saw in the technical best practices how handling the validity domain and the model accuracy were important to produce good results. From an outside point of view, these details are not considered, and the results will be used whether accurate or not. Wrong decisions can be made and a model can lose credibility because of this. Teams must legitimately try to stop this kind of excess.
  - *A bad reason*: It can be the engineers' feeling of owning a model; they spent time and put the best of themselves into it. They can be frustrated to see their "baby" going away, to see other stakeholders using it and taking the credit for efforts made by others (especially in companies where simulation is underestimated, as mentioned before). In this case, the simulation engineers who built the model may try to avoid losing control of it. Wrong but credible arguments can be brought up, just like in the good reason above (a situation where this is unlikely), or potential model exchange–related technical issues may arise that could easily be fixed.
- *Setting a climate of trust*: To prevent these two situations above, a climate of trust is mandatory. This can come from two unbreakable rules (this can fall to the numerical simulation pilot, the simulation manager, or even the simulation senior):

1. *Monitoring the models use*: This is about warning the stakeholders about the potential threats we mentioned previously (wrongly using a model). This use is often unintentional and communicating about it is enough to prevent it. In some rare cases where this drifting occurs, some appropriate measures can be taken in order to raise awareness of this rule among all stakeholders (keeping in mind that this occurrence is probably linked to a misunderstanding of this rule and the one supposed to broadcast this rule is responsible!).

2. *Involving the model designers in results communication*: The purpose is, first, to make sure the model is properly used, and second, to value the engineers who built the model (as the model designers are scarcely able to introduce their work, which is a source of frustration).

Communicating on these two rules and ensuring their application enables one to set a climate of trust that stimulates a positive state of mind around the model exchanges within the company.

5.2.7.3 Virtues of Open Models

The sources of potential issues regarding model exchanges and their solutions have just been covered. As one last point, being able to fix most of the issues mentioned previously isn't less important, but some notions need explaining.

It is about open models. What are these? These are models that can be fully explored, of which no area is encrypted (the opposite of closed models, of which the black boxes mentioned before are an example). The models exchanged internally at least need to be open for the following reasons:

- *Minimizing IT problems*: These can be caused by black boxes, as we detailed before; keeping the models open prevents that situation.
- *Minimizing the consequences of miscommunication*: For instance, when the model is open, checking the input/output data is made easier by exploring the inside of the model.
- *Easing the model controlling*: An open model is easier to apprehend, as its content can be briefly explored. Also, when bugs happen, fixing is simplified by this opportunity.
- *Conveying knowledge*: Open models lead to a better broadcast of the knowledge within the company. The models include the knowledge; open models make it visible.
- *Favoring innovation and creativity*: The model's parameters often match the designing team's parameters. Assembling open models

offers a more natural and global vision of the systems, caring less about the design team's parameter interfaces (since the model is no longer just an assembly of black boxes related to their interface). Simulation often brings new ideas, transverse, that previous interfaces were inhibiting.

Thus, for all the reasons mentioned previously, the models exchanged need to be open. Some obstacles occur, such as the state of mind we mentioned that must be dealt with by setting a climate of trust. An argument against open models is that they can be a source of problems as they are editable by the user, who might be inexperienced. But many software programs enable one to keep the model open and block any changes.

So, setting a climate of trust will enable an open-model culture within the company, boosting exchanges, and, in this way, expanding numerical simulation.

APPLICATION OF BEST PRACTICE 7

The synthesis modeling team's numerical simulation pilot became aware of the previous best practices. Within a few weeks, he managed to apply the following solutions:

- *The team modeling the internal thermics*: He decided to set up a test case system (as well as for the other teams, for that matter). So, each time a model was received, if it didn't work or if the results did not comply with the ones expected, the team could entirely focus on problem-solving this case, making the adjustments much more efficient.

- *The team modeling the cooling loop*: After referring to the project stakeholders, the privacy could be removed. The cooling loop–modeling team just needed to encrypt the models; fixing the model's few errors was then much more simple.

- *The team modeling the command control*: The simulation pilot devoted the supplier's time to setting up the model input and output interfaces. Thus, the synthesis modeling team saved the energy of making adjustments. Moreover, the supplier stated he also saved time thanks to that standardization: it was then easier for him to compare the different historical model versions.

So, by better handling model exchanges and distribution, the company managed to improve the efficiency of its simulation teams.

5.2.8 Best Practice 8: Managing Configurations of Models and Results

MISTAKE 8

Back to the modeling team: Numerical simulation kept on expanding, so much so that a great simulation campaign was run in order to compare different potential fridge designs.

Project manager: "You sent me simulation results 2 months ago. I compared them with those you sent me this morning. I'm surprised, because the electrical consumption from this morning's studies are higher than those from 2 months ago, when I was expecting the opposite."

The team modeling engineer finally recovered the results sent to the supervisor 2 months beforehand. Still, he had three questions about these 2-month-old results.

1. To which (fridge) design hypotheses were the results related? Some of the hypotheses have changed in the meantime, and the isolation of the fridge has potentially increased; were these changes included in the model? He couldn't remember.

2. With which level of model maturity do these results correspond? Some changes were performed on the model to increase its accuracy. This changed the results a little. But was it before or after these 2 months?

3. To which life situation were these results related? A standard life situation change was defined (the door openings were not occurring at the same time anymore). Was this change included in the results?

Failing to answer these questions, the simulation engineer couldn't identify the source of these gaps between the current results and those from 2 months ago. He couldn't say whether these gaps resulted from a mistake or if they revealed some evolutions. This brought suspicion over all the simulation campaign results and dramatically limited the impact of the results on the project.

So, by failing to manage the models and results configuration, the company didn't manage to benefit from all the contributions of numerical simulation.

In order to properly expand numerical simulation, managing the model configurations and results is crucial, and they multiply very fast. Identifying which model some results come from (e.g., with what accuracy) and what systems they refer to (e.g., with what pieces definition within the system) can be impossible. Without model settings and results management, these can become absolutely useless and lead to poor decisions.

Why are model configurations and results management that complex? This complexity has three sources of diversity:

1. *Depicted system diversity*: The system, represented by the model, generates itself a wide range of potential configurations. A complex system often includes pieces (organs), control commands, and gradings. These three components gradually change during the model design. A project may be considering several pieces, command controls, and grading opportunities for the same system. The number of system configurations can shoot up and so can the models depicting this system.

2. *Modeling type diversity*: The model itself has different configurations for the same system. First, with several types of model (e.g., with different accuracy levels). Second, when the modeling work progresses, maturity increases, and the model matures (but not the system, which doesn't change in configuration). Finally, the software can change (e.g., it can be updated), and the models follow these changes to remain compatible. So, after a few months' work modeling a system, even if it doesn't change, the number of models has often blown up.

3. *Simulation results diversity*: Finally, several results can be generated with the same model. For instance, different life situations will be tested, several configurations will be used, or even different output data will be recorded. Connecting simulation results with their model can quickly become complex if precautions were not taken upstream.

These were the three sources of diversity complexifying the configuration management work, making it crucial. Now let's see how to run these configurations. Three steps allow us to do that.

5.2.8.1 Raising Awareness of Configuration Management

Raising awareness of the complexity and the stakes of configuration management was the previous section topic. The saying "Forewarned is forearmed" makes a lot of sense here. Being aware of the previous points makes the team more cautious. Even if no specific tools are available to manage these configurations, anyone aware of both complexity and importance will be able to reduce mistakes (by storing both models and results in an organized way).

5.2.8.2 Using a Tool

Many software programs allow one to manage model configurations and results (or at least they help).

- *Necessity of a tool*: Managing configurations without a tool is theoretically possible (e.g., by noting data storage locations and the connections of this information for baseline documents). Still, as the number of data can be consequential, proceeding with management without any tools is not an easy task (it would be both complex and time-consuming). Using a tool is wiser.
- *Picking a tool*: Just as with modeling/simulation software (Best Practice 4), picking configuration management software requires one to cross-reference needs and offers to assess the most relevant ones.
  - *Identifying the need*: Some points need to be broached to perform this identification.
    - *Complexity level handled by the software*: How complex can the company configuration management case be (especially the system diversity, the modeling type, the results)? But also, how much assistance from the software do the stakeholders expect?
    - *Relation with system configuration management*: In order to link the model configuration and the system configuration, the latter must obviously be managed too. In practice it hardly is, and sometimes no tool is used at all. However, the software managing the model configurations needs to be compatible with the software managing the system configurations (if it exists). Using both these software programs in a synchronized way may be appropriate (or maybe using the same software).
    - *Storing and/or referencing*: One must determine whether the software's purpose is just to manage the data referencing (building connections between them) or also to manage its storage.
    - *Other factors*: Besides these points, some other factors will be considered, such as the software ergonomics, the support, the costs, and the long-term reliability. The reader may refer to Technical Best Practice 4, where some points are detailed when the software choice remains applicable.
  - *Identifying the potential offers*: The range of software offering to manage model configurations is tighter still than that of simulation software. Options will be limited in this case (compared with the situation described in Best Practice 4).
    - *There is no perfect tool*: Over the last few years, the tools allowing one to manage the data lifecycle (product lifecycle management; PLM) has boomed. The ambition of these tools is to manage all product (system)-related information and data during their entire lifecycles. Simulation models and results belong to one of these. In practice, no tool is perfectly

satisfying and none of them really stand out. The need has to be identified (refer above) to know what points compromises can be made on.

- *Diversity of marketed tools*: There is a wide range of potential tools. Some of them are very simple (open source), while some others are much more all-encompassing. A simple and flexible tool is often much more efficient than a rigid and complex one.

5.2.8.3 Efficient Use of That Tool

We just covered how to properly choose numerical simulation results and model configuration management tools. Possessing a tool is one thing, being able to correctly use it is another.

- *Stakes of efficient use*: If the tool isn't properly used, efforts will have been made for nothing. A complex tool is often chosen, while the teams do not use it to manage their data configurations or use it inappropriately, to the extent that it becomes useless.
- *Driving configuration management*: Having a stakeholder (often the simulation driver or potentially a configuration management stakeholder) who makes sure the model exchanges and storage are done in consideration of the configuration management best practices, and who knows how to use the appropriate tool, is important. This may seem daunting at the moment (as this step doesn't allow value creation but only prevents value losses), but the few minutes invested often save many hours or days of extra work.
- *Future use of the tool*: This is to be taken into account in the tool choice. An exploited basic software program is worth more than a unused complex one.

These best practices should help the reader properly manage the models and results in configuration. Finally, two examples can be listed to take a little perspective: one where the configuration is at a good level (evidence that reaching a good level isn't impossible) and another one where configuration is low (evidence that simulation isn't the only thing that may not be properly managed).

- *A good example of configuration management*: Nowadays, product nomenclature management has reached a very high level of maturity and product configurations (which can be very varied and complex) to be manufactured are very well managed.
- *A shaded configuration management example*: Live experiment configuration management is often perfectible. First, live experiments are often performed with means lacking the expected configuration (for lack of

traceability and coordination). Second, these live experiment results are often poorly managed in configuration, to the extent that conveying with which means the experiment was performed, corresponding to what system version, becomes impossible, making the result useless.

APPLICATION OF BEST PRACTICE 8

Let's go back a few months earlier. The company became aware of the previous best practices, and the simulation manager decided to set up a model configuration and numerical simulation results management tool. The one he opted for wasn't perfect, but it enabled him to prevent the main inconveniences of a lack of configuration management. He also named the simulation driver who would be in charge of making sure this tool was used by the teams.

Once again, the project manager asked the simulation team to explain to him the gaps between the results he was sent 2 months ago and those he received that morning.

This time, the simulation engineer found the former result in the configuration management tool and was able to compare it with the last one.

- *Represented system diversity*: The fridge design hypotheses were the same.
- *Modeling type diversity*: The modeling type wasn't the same. There was a model modification in the meantime to increase the accuracy, which had a slight impact on the results. This modification increased the electrical consumption value a little.
- *Simulation results diversity*: The results type corresponded to the same situation (especially the same life situation studied and the same configuration).

The simulation engineer was able to answer the project manager.

Simulation engineer: "The results gap comes from a model modification to increase its accuracy. This rise of electrical consumption isn't intrinsic to the system, but it is related to another way of depicting it."

The project manager took note of these gaps to proceed with his comparisons, and the simulation team results seemed credible to him.

So, by properly handling the numerical simulation results and the model configuration management, the company managed to maximize the benefits of the simulation.

6

The Future of Numerical Simulation in Industry: The Early Twenty-First Century

Numerical simulation hasn't reached its maturity yet and remains fast growing. The changes formerly initiated haven't been entirely broadcast, and more transformations are still to come, encouraged by IT tool evolutions, simulation techniques, and industry.

Forecasting enables us to drive these changes or at least to anticipate and adapt.

6.1 Why Is Predicting the Future Important?

This chapter's objective is to offer forecasts of the changes to come regarding numerical simulation. Why? To make the reader a driver of these changes or at least enabling them to anticipate and adapt. Our goal isn't to speculate but to help the reader efficiently build that future. Of course, the vision we will offer isn't exhaustive, and potential disruptions that can't be foreseen might occur.

We are going to review the numerical simulation chronicles covered in Chapter 2 and carry them on predictively.

We will focus on the next 20 years (forecasting beyond that point is seldom useful in industry) and briefly mention longer-term perspectives at the end of the chapter.

6.2 The Next 20 Years

We will mainly focus on the upcoming changes for the next two decades, starting with those initiated previously that aren't fully expanded today and broach those we expect in the next few years. We will finally see their overall impact on numerical simulation condition.

The scope of our study only covers numerical simulation and doesn't deal with the IT field (which is much larger and in which changes are more quantitative).

6.2.1 Changes from the Past, Still in Progress

Numerical simulation is first an activity that hasn't yet reached its maturity, in the sense that its expansion is still ongoing. So, even if no innovations occur later, numerical simulation will keep on expanding. A great number of companies or fields haven't yet opted for numerical simulation though it could be profitable. This delay can be explained by multiple factors, among which are the distribution of perfectible knowledge or the inertia of organizations. Among these situations, expansion is just a matter of time.

This first point we just mentioned is enough to predict the positive growth of numerical simulation activities for the next few years.

6.2.2 Changes to Come

6.2.2.1 Introduction

We are going to focus on the changes to come. They will impact on numerical simulation expansion in addition to the previous point.

We explained that numerical simulation uses *numerical tools* and *simulation techniques* to serve *industry*. So, three main kinds of changes related to these three fields can occur:

1. *Numerical tools*: The changes related to this domain open up a realm of possibility and enable one to widen the simulation activities. For instance, as mentioned in Chapter 2, the expansion of personal computers in the 1980s greatly helped to broaden numerical simulation.

2. *Simulation techniques*: The changes can also impact on numerical expansion. They are frequently triggered by other field changes (e.g., numerical tools or industry progress). Still, we will mention here changes that always include some specificities and innovations related to simulation techniques, so much so that simulation techniques have become fields in their own right. For instance, although it is true that computer-aided design (CAD; see Chapter 2) was triggered by the emergence of computers, it mainly expanded independently, going through major innovations specific to it that were not provoked by another field.

3. *Industry*: Industry changes also have a strong impression on numerical simulation expansion. They can provide numerical simulation with new fields of application, or numerical simulation has to adapt to them. For instance, over the last decade, driving assistance in the car industry has encouraged a significant demand for simulation (to prevent the building of a large number of prototypes and ensuring the robustness of the embedded software).

We have identified the three main fields, mostly distinctive, in which changes will impact numerical simulation and define its future. The reader will notice that two of these factors are external to numerical situations (numerical tools and industry) and one is internal (simulation techniques). Even the external factors are partly internally connected. For instance, numerical simulation techniques, via their needs for numerical tools, have stimulated numerical tool growth. So, the internal and external influence each other. Still, even if that interaction exists, a frontier remains.

We will now cover the change triggers among these three fields that will impact numerical simulation activity, starting with both the internal domains and ending with the external. We will go further regarding the internal change triggers than with the external. Note that our goal isn't to list all changes in the three fields, but only the one impacting numerical simulation. These changes are depicted in Figure 6.1, and we will now detail them.

6.2.2.2 Change Catalysts Linked to Numerical Tools

What are the future numerical tool evolutions that will impact numerical simulation? Three are discussed here.

6.2.2.2.1 Explosion in Available Information

- *The change*: Following NTIC progress, collecting large amounts of data has become simpler. New techniques have been developed and continue to be developed to store and process these data (big data), as well as to read and use them (data science, data mining, machine learning). Data explosions, in addition to connected techniques, are probably going to provoke a burst in the next few years, thus a new major disruption in the numerical field.
- *The impact*: Numerical simulation is very data-consuming. These data allow one to configure, define, and validate a model. An explosion in the available data clearly offers great opportunities to numerical simulation and will probably allow cheaper designs of models to be made. Nevertheless, although a data explosion may provoke a disruption in personal lifestyles, the disruption in the area of numerical simulation is likely to be lower. In the same way, if the Internet had a strong impact on numerical simulation 20 years ago, its effect was much less on personal lifestyles.

6.2.2.2.2 Increase in the Computing Power Available

- *The change*: In the past, the increase in the calculation power of computers was exponential, and it is incontestably going to carry on as such. Thus, besides the increased calculation power of each computer, it is now easier to connect several computers, and in this way to have more computing power available (today, by counting up the

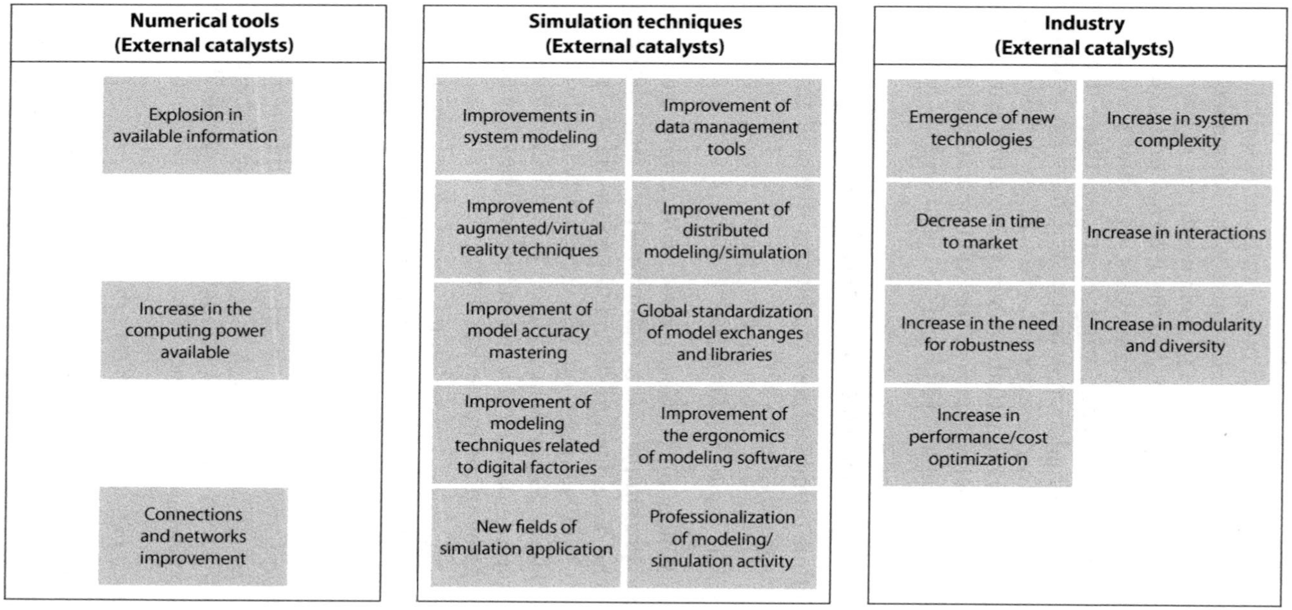

Numerical tools (External catalysts)		Simulation techniques (External catalysts)		Industry (External catalysts)	
Explosion in available information		Improvements in system modeling	Improvement of data management tools	Emergence of new technologies	Increase in system complexity
Increase in the computing power available		Improvement of augmented/virtual reality techniques	Improvement of distributed modeling/simulation	Decrease in time to market	Increase in interactions
Connections and networks improvement		Improvement of model accuracy mastering	Global standardization of model exchanges and libraries	Increase in the need for robustness	Increase in modularity and diversity
		Improvement of modeling techniques related to digital factories	Improvement of the ergonomics of modeling software	Increase in performance/cost optimization	
		New fields of simulation application	Professionalization of modeling/ simulation activity		

FIGURE 6.1
Three types of change triggers benefiting numerical simulation expansion.

average time computers are not used, we can assess that 98% of computing power is not used). If it is undeniable that computing power is going to increase, the question regards the amplitude and the pace of that increase (new technologies are likely to create disruptions in that field).

- *The impact*: The increase in calculation power available obviously helps numerical simulation expansion. Indeed, many simulations still remain too demanding of computing power to be performed (e.g., in the field of research). Then, less calculation time supports the simulation of a larger amount of models in real time and broadens the realm of possibility. That impact looks unlikely to cause any major disruption in the numerical simulation field (except in the case of a major disruption in calculation power).

6.2.2.2.3 Connections and Networks Improvement

- *The change*: The Internet is now mature after more than 20 years of existence, but innovations in networks and connection areas still occur (e.g., connected devices or gains in communication speed). Computer tools are more and more interlinked, and the frontier between one computer and another tends to shrink.

- *The impact*: The continuous improvement of networks and connections will allow the efficient exchange of models (whether internally or externally—e.g., with suppliers). Also, remote calculations (when the modeling and/or simulation work is performed on a remote computer) require tough networks and connections with remote servers. If no major change is scheduled, this IT progress will ease the expansion of simulation.

We have covered three change catalysts in the area of numerical tools, promising a positive impact on numerical simulation expansion.

6.2.2.3 Industry-Related Catalysts

We will now focus on the industry evolutions with a consequence on numerical simulation. As mentioned before, the points to be covered here are theoretically external to numerical simulation but are often partly influenced by it.

Which future evolutions in industry will have an impact on numerical simulation? Here we mention seven of them.

6.2.2.3.1 Emergence of New Technologies

- *The change*: In the upcoming years, new technologies will grow, creating new industrial sectors.

- *The impact*: The growth of industrial activities will sustain the growth of simulation activities. Indeed, these new technological developments will often require simulation activities. For instance, studies on self-driving cars in the car industry call upon simulation. First, hypothesis feasibilities need to be assessed very early in the upstream design process, and second, simulation is essential to guarantee the robustness of features in very varied life situations (a multitude can be tested, which is impossible in practice). So the rise of new technologies benefits the expansion and renewal of numerical simulation.

6.2.2.3.2 Increase in System Complexity

- *The change*: Industries will develop more and more complex systems to better fulfill the consumers' and customers' needs. For instance, aircrafts have become much more complex than 10 years ago to adjust for environmental and financial constraints as well as standards changes.

- *The impact*: Numerical simulation is a great tool to manage complexity, as a model is able to include much more physical phenomena and data than a human brain. So, the systems complexity increase is a growth potential for numerical simulation.

6.2.2.3.3 Decrease in Time to Market

- *The change*: Companies will always expect to shorten their time to market as much as possible, whether it is to get closer to the expectations of consumers (who may rely on time, and who may not be the same at the beginning and at the end of the product design process) or to be innovation leaders, or again to minimize the negative impact of immobility on profitability.

- *Impact*: We saw in Chapter 3 that numerical simulation is especially relevant to decreasing the time to market. This tendency is a growth source for simulation.

6.2.2.3.4 Increase in Interactions

- *The change*: In a world of globalization, with goods that are more and more complex, an increasing number of stakeholders need to work together, and this tendency will go on.

- *Impact*: We saw in Chapter 3 that simulation enables us to communicate and work together efficiently (the model then includes the state of the art of conception). Thus, numerical simulation will allow us to answer that need for increased interactions and will contribute to its expansion.

6.2.2.3.5 Increase in the Need for Robustness

- *The change*: The requirements in terms of the predictability and reliability of products will keep on increasing. This need comes from the increases of both constraints in terms of security and our standards of living.
- *Impact*: We mentioned in Chapter 3 that simulation is an efficient tool to improve robustness. For instance, the aircraft and engine design processes are often assisted by simulation to guarantee correct operation in many situations.

6.2.2.3.6 Increase in Modularity and Diversity

- *The change*: First, the expected diversity of consumers and clients will keep on increasing (for the marketed system to fit their expectations). Second, the modularity within companies will grow as well, to handle the explosion in systems diversity (succeeding in generating a wide diversity of module combinations).
- *Impact*: We saw in Chapter 3 that simulation is relevant to supporting a modular process. This tendency for modularity increases is profitable for numerical simulation.

6.2.2.3.7 Increase in Performance/Cost Optimization

- *The change*: In mature fields, strong constraints (especially competitive) lead to a requirement for system optimization.
- *Impact*: As mentioned in Chapter 3, simulation is a great tool for optimization, and this is also profitable for its expansion.

6.2.2.4 Simulation Techniques–Related Change Catalysts

We will now approach the evolution of simulation techniques, which will impact their expansion. So, the changes we are about to see are internal. They were or will often be triggered by external constraints or opportunities. Nevertheless, here we mention some specific innovations. Ten main catalysts of this type can be cited.

6.2.2.4.1 Improvements in System Modeling

The purpose of modeling a system is to guide complex systems design. System modeling enables one to conceive the system command control and pre-size the different parts. It often calls on 0D and functional modeling techniques (see Chapter 4). Because of the changes related to industry we mentioned before (especially the increase in the complexity of systems), new techniques have grown in the system modeling field, and will keep on doing so in the upcoming years. Here are the most important:

- *Demand modeling/simulation*: Nowadays, system specifications are almost systematically written down. Some techniques have emerged to model these demands and make them experienceable: these techniques (including the associated modeling languages) will keep on evolving. This will improve both the efficiency and the continuity of the design process (the system conceived can be tested and approved through a comparison with the demand behavior).

- *Multiphysical and team modeling*: As systems become more and more complex, many varied stakeholders need to be involved as well as varied physical domains. For a complex system to be modeled, several models including different physical phenomena designed by several stakeholders need to be involved (internal or external to the company). This raises some technical and organizational issues. Over the last few years, lots of progress has already been made. For instance, many co-simulation software programs have come out (allowing one to pair different types of software), and the editors have managed to build paths between their software.

- *Automatic code generation*: Embedded codes in systems are still usually manually encoded. Several situations can occur. Command controllers can be considered, then directly encoded manually. In some other cases, as surprising as it may seem, the command controllers are modeled, then written down, and then manually encoded afterward in the embedded systems. These cases occur when there is no proper design continuity (e.g., if the design stakeholders are displayed in distant teams from an organizational point of view or even if a part is subcontracted). Some software programs enable automated code generation, coming from the model that will be implantable in the system's embedded software. Progress is still to be expected in that field for the code-automated generation to no longer be a best practice but an obvious one.

- *Design validation and control techniques*: Progress is also to be expected to ensure the consistency of system design (the goal here isn't whether to check/approve the model but the system itself). A system model includes the intrinsic mechanisms explaining its behavior (the system is either a control or a physical part). Theoretically, one can identify which requests to type as the system input, to exhaustively (and as efficiently as possible) monitor the system's behaviors (according to the specification model). For instance, in the case of command control, some software creates life situations allowing one to test all the insight branches of the command controller. Thus, these life situations can be experienced in the live system (embedded software) to make sure it behaves as expected. This offers great opportunities to improve the system's robustness, and progress is also to be expected here.

- *Mastery and continuity of the MIL/SIL/HIL*: Let's start by defining these three notions related to the design of system command controllers:

    - *Model in the loop* (MIL): Consists in testing the command controller model by pairing it with an environment model (the environment model depicts the model itself without the command controller, just as the outside environment). This enables one to assess if the command controller model meets the expectations.

    - *Software in the loop* (SIL): Consists in testing the command controller code (based on the command controller model) in an environment model. This enables one to check if the code transcription is correct.

    - *Hardware in the loop* (HIL): Consists in testing the embedded software (in hardware versions) in an environment model. This enables one to make sure the integration in the embedded platform hasn't provoked any distortion.

    - *Change*: These techniques are currently spread throughout industry and keep on evolving, one of the challenges being to understand the continuity of these works. It will guarantee the robustness of the embedded command controllers and minimize their associated costs.

- *Toward a global functional model*: The digital model was born a little more than 20 years ago (see Chapter 2) and included a 3D representation of the whole considered system, even before the first system prototype existed. A functional model is likely to appear in the next few years; it will include a functional representation of the system studied, meaning its global behavior in any life situation will be able to be predicted. There are still some restrictions (like continuity mastery or the opportunity to easily connect the different models constituting the systems), but they will disappear eventually.

6.2.2.4.2 Improvement of Data Management Tools

We saw in Chapter 5, in Best Practice 8, that there is a major stake in managing to efficiently handle model and simulation settings. This need is escalating, especially regarding the simulation nonrelated context evolution. There are many marketed software programs today, especially some that integrate wider features to cover product lifecycle–related information management (product lifecycle management; PLM).

This software is still being improved and it is not yet ready (and has not so far expanded much in industry or with only limited features).

6.2.2.4.3 *Improvement of Augmented/Virtual Reality Techniques*

We saw in Chapter 2 that virtual reality really started entering industry in the late 1990s. Nevertheless, today, while it could bring a lot to industry, its applications are still limited. Indeed, there are still many barriers, mostly concerning the field of numerical tools (computing power) and the intrinsic techniques of virtual reality. These barriers will open continuously, opening up a realm of possibility.

6.2.2.4.4 *Improvement of Distributed Modeling/Simulation*

Although these techniques are 20 years old, they are still common. The current limitations are related to numerical tools (transfer speed on networks) and their inner techniques. With the growth of numerical simulation and model exchange activities, the need for simulation and model transfers onto dedicated remote servers will increase. The distributed modeling/simulation techniques will grow, and working on remote computer-managed models will become more and more common, as well as remote simulation.

6.2.2.4.5 *Improvement of Model Accuracy Mastering*

We have seen throughout this book that mastering model accuracy is a complicated matter; still, it is essential. That difficulty will increase with the emergence of more and more complex models (enabled by powerful numerical tools and a growing need for the mastery of system complexity). For instance, mastering the accuracy of a collective multiphysics model, in which many stakeholders are involved, isn't an easy task. Some techniques will emerge to succeed in mastering the accuracy of such models, and the latter should improve. Also, some rules are emerging whose purpose is to help communicate the level of accuracy of models, in a standardized way.

6.2.2.4.6 *Global Standardization of Model Exchanges and Libraries*

We saw in Chapter 5, Best Practices 5 and 6, how important it is to succeed in exchanging the models within a company and avoiding the replications. This same remark makes all the sense when it comes to exchanging the models in the rest of the world, outside the company (especially among companies, universities, and organizations). Three main changes should improve the distribution and exchange of models around the world.

1. *Model exchange standards*: There are no world baseline standards for exchanging models. There are several standards (none has taken the advantage), and each of them is often associated with one single editor, which complexifies their distribution. So, exchanging the models hasn't yet become a simple task. This is likely to change in the future.

2. *Models connectivity*: As we mentioned before, connecting the models isn't simple yet. This requires technical evolutions, partly on the

editors' side. Also, connecting different models together is likely to become much easier soon.

3. *Model libraries*: Guessing the amount of potential model replications worldwide would give one a headache. Let's consider an alternator: How many people in the world have built an alternator model? As it is a common part, probably over 10,000 alternator models have already been built. Thus, thousands of people have worked on models that already existed. That diversity doesn't bring any value, but it has a cost (the reader may object that this diversity enables several levels of modeling precision, but that diversity could only justify about 10 of them). So, we are in the same situation as 30 years ago, when simulation software editors figured out that many modelers were re-encoding, in duplications, the same numerical features. Softwares (that are a references today) at this time were born with pre-encoded numerical features, saving the re-encoding time and improving the quality of these features (for it was coded by experts). The same situation is very likely to come up in the years to come, regarding physical system models this time. This change seems to have started (more and more editors offer libraries) but hasn't yet been on the rise (partly for the reasons mentioned previously).

6.2.2.4.7 Improvement of Modeling Techniques Related to Digital Factories

The digital factory is a notion consisting in using numerical innovations to profit industrial production. This is a very broad notion that was born simultaneously with numerical simulation, little more than 60 years ago. Digital factories call on simulation (but not exclusively). Among the simulation application examples in the digital factory field, we can mention the design of assembly lines, automated dynamic optimization, operator training, activities scheduling, or stock management and production forecasts. Simulation is already helping industrial production. Nevertheless, on the one hand, these technologies keep on upgrading and renewing. On the other hand, industrial production has to face new challenges: dealing with increasing product diversity, more complex products, and guaranteeing better reliability, while always keeping costs low. To take up this challenge, new progress in digital factories is required. Thus, digital factory–related modeling techniques will probably keep on upgrading.

6.2.2.4.8 Improvement of the Ergonomics of Modeling Software

We have seen that, throughout history, modeling software has become more and more ergonomic, aiming especially at simulation juniors. Simulation will keep on expanding, and so it will be used more and more by nonspecialists. Demand for easy-to-use tools will also keep on growing, and this tendency is very likely to persist in the next few years.

6.2.2.4.9 *New Fields of Simulation Application*

Thanks to the growth of simulation, it will be able to expand over new fields (and already-existing fields but in which simulation is hardly used or unused). Some of them have already been mentioned (e.g., digital factories), but the scope of opportunities is broader than the few examples we have seen. Simulation is very likely to keep on expanding in new fields such as health care, company running, or project monitoring, and these will require the development of new dedicated techniques (here it is about new applications of simulation, but in already-existing domains, contrary to the points broached in industry-related change catalysts, where the fields in question do not yet exist).

6.2.2.4.10 *Professionalization of Modeling/Simulation Activity*

We saw in Chapter 5, in Best Practice 5, that industrializing numerical simulation was required once the activity proved its worth. We will go further into this point here. Numerical simulation will keep on expanding in industry, it will become crucial for best practices and techniques that were isolated to become massively expanded. It is now time for the simulation activities to be professionalized. One could state that it has already been done in some parts of numerical simulation in some companies, and that would be right. Nevertheless, the expansion isn't yet massive. The best practices we have covered are likely to expand on the company side; companies will have simulations industrialized, put them into projects, and apply *lean* practices. On the education side, numerical simulation will probably—and this is desirable—have increasing importance in the syllabus (the contents of training will have to adapt according to the role chosen by the students).

We have covered the main changes to come within three fields: numerical tools, industry, and simulation techniques. These changes will impact numerical simulation.

6.2.2.5 *Consequences of These Changes*

We just saw that in the numerical simulation area, former changes are still occurring and new changes are to come. What are the consequences?

The main consequence is that numerical simulation will keep on expanding in industry and that the growth period (relatively uninterrupted for the last 60 years, although the pace wasn't constant) isn't over. This prediction of ours regards the next 20 years. So, numerical simulation will keep on expanding in some fields over new issues but also appear in some others where it hasn't yet been used. A second consequence is that numerical simulation will renew again as new techniques will replace the former ones.

What will the pace of this growth be? What will the scope of the numerical simulation expanding over the next 20 years be? We saw that a technical innovation was unlikely to provoke a major disruption in the simulation field during this period (despite foreseeing true growth). Still, a massive expansion could occur for the organizational purpose we mentioned before. By observing the past, we can note that the highest pace was during 1980–1995 (see Chapter 2), a period that profited from the expansion of personal computers. The following period (the last 20 years) has seen smaller growth. In the end, the next 20 years' growth is likely to be between the two others. Defining it by the growth of the profits generated by simulation in industry, this growth is likely to be in the double digits.

What lesson should be learned? First, one must be aware of the growth of numerical simulation and the value creation it brings, using it as an opportunity (taking place as a leader in the simulation field) and not as a threat (watching the competitors taking advantage of it and taking the lead). Secondly, we provided our vision of these growth ingredients, for each stakeholder to contribute and be a driver.

Finally, let's make three remarks:

1. *From CAD to design performed on computers*: Since numerical simulation has entered industry, the common vision is the one of CAD. Simulation is seen as an additional tool. This vision will probably and slowly reverse for the reasons mentioned previously. It will become obvious, and conversely, designing a complex system without modeling will be very surprising. (Since a computer is able to master greater complexity than a human brain, if the right precautions are taken, these decision will be much more relevant than those of a human.) Numerical simulation will become central in the design process, and we can assess the simulation won't be *computer aided*, but *computer performed* or *computer decided*.

2. *Non-exhaustivity and imperfection of the prediction*: A prediction is obviously imperfect and non-exhaustive; some elements of this chapter are likely to be contradicted by reality in the future.

3. *Design without live experiments*: Design in the future is commonly said to be free of live experiments. Prototype experiments are likely to disappear on the long term. Nevertheless, this doesn't mean experiments won't be necessary anymore. Understanding systems (and subsystems) will remain crucial. However, these live experiments are likely to be performed more on subsystems in an isolated and accurate way. Characterizing the system won't be necessary anymore, as its behavior is nothing but the assembly of the behaviors of previously characterized subsystems.

6.3 And Then?

We've tried to provide a vision of the potential changes to come on a 20-year scale, for the reader to be the driver of these changes and/or to adapt to them.

After this 20-year period, numerical simulation is likely to dramatically mutate and expand. Nevertheless, forecasts beyond 20 years are hardly accurate and often useless (at least in industry). So, we won't go any further in forecasts, but on the contrary, we suggest two questions:

1. *What role will simulation languages have?* Simulation languages are particularly efficient at explaining a system behavior (and capitalizing the understanding). These representations are denser, more efficient, clearer, and more intuitive than commonly spoken languages. Could simulation languages, in the technical fields, replace spoken language?

2. *What will be the place for humans in industry?* Today, simulation has already replaced and canceled some human-managed activities. Little by little, simulation is taking an increasingly large place, even in fields where it wasn't expected. It is possible to "create" ideas, to innovate, through simulation, but also to make company management decisions, project monitoring (simplifying the decision-making process, learning from the past). Already in some situations, the task of humans is only to define the objective of the study, which is then performed by a computer. What won't a computer be able to do that a human will?

Conclusion

We have seen that numerical simulation will keep on expanding and mutating in the upcoming years and that poor knowledge sharing hinders these changes.

This book's purpose is to help that broadcasting of knowledge, and we hope we have answered the questions raised in the introduction, whether it is for directors, project managers, simulation stakeholders, or students and teachers.

The responsibility falls to the readers to put the best practices we have seen into practice in the best way they can.

Finally, your author always accepts, with much interest, any comment from his readers at the address bookfeedback.simulation@gmail.com.

Bibliography

Balci O. 1990. Guidelines for successful simulation studies. In *Proceedings of the 1990 Winter Simulation Conference*. Piscataway, NJ: IEEE.

Balci O. 1994. Validation, verification, and testing techniques throughout the life cycle of a simulation study. In *Proceedings of the 1994 Winter Simulation Conference*. Piscataway, NJ: IEEE.

Banks J., Carson II J. S., and Nelson B. L. 1996. *Discrete-Event System Simulation*, 2nd edn. Englewood Cliffs, NJ: Prentice Hall.

Breno B., de França N., and Horta Travassos G. 2013. Reporting guidelines for simulation-based studies in software engineering. *Evaluation and Assessment in Software Engineering (EASE 2012)*, 16th International Conference, Ciudad Real, Spain.

Chwif L., Barretto M. R. P., and Paul R. J. 2000. On simulation model complexity. In *Proceedings of the 2000 Winter Simulation Conference*. Piscataway, NJ: IEEE.

Emshoff J. R. and Sisson R. L. 1970. *Design and Use of Computer Simulation Models*. New York, NY: Macmillan.

Fishwick P. A. 1995. *Simulation Model Design and Execution: Building Digital Worlds*. Englewood Cliffs, NJ: Prentice-Hall.

Fritzon P. 2014. *Principles of Object Oriented Modeling and Simulation with Modelica*. Hoboken, NJ: Wiley.

Hills P. R. 1971. *HOCUS*. Egham, Surrey, UK: P-E Group.

Hollocks B. W. 2008. Intelligence, innovation and integrity, KD Tocher and the dawn of simulation. *Journal of Simulation* 2(3): 128–137.

Innis G. and Rexstad E. 1983. Simulation model simplification techniques. *SIMULATION* 41: 7–15.

Kleijnen J. P. C. 1995. Verification and validation of simulation models. *European Journal of Operational Research* 82: 145–162.

Kotter J. P. 1996. *Leading Change*. Boston, MA: Harvard Business School Press.

Law A. M. and McComas M. G. 1991. Secrets of successful simulation studies. In *Proceedings of the 1991 Winter Simulation Conference*, ed. Charnes J. M., Morrice D. M., Brunner D. T., Swain J. J., pp. 21–27. Piscataway, NJ: IEEE.

Ljung L. and Glac T. 1994. *Modeling of Dynamic Systems*. Englewood Cliffs, NJ: Prentice Hall.

Nance R. E. 1993. A history of discrete event simulation programming languages. The Second ACM SIGPLAN Conference on History of Programming Languages, April 20–23, 1993, Cambridge, MA, pp. 149–175.

Nance R. E. and Sargent R. G. 2002. Perspectives on the evolution of simulation. *Operations Research* 50(1): 161–172.

Nance R. E., Goldsman D., and Wilson J.R. 2010. A brief history of simulation revisited. In *Proceedings of the 2010 Winter Simulation Conference*. Piscataway, NJ: IEEE.

Naylor T. H., Balintfy J. L., Burdick D. S., and Chu K. 1966. *Computer Simulation Techniques*. New York, NY: Wiley.

Oscarsson J. and Urenda Moris M. 2002. Documentation of discrete event simulation models for manufacturing system life cycle simulation. In *Proceedings of the 2002 Winter Simulation Conference*. Piscataway, NJ: IEEE.

Richter H. and Marz, L. 2000. Toward a standard process: The use of UML for designing simulation models. In *Proceedings of the 2000 Winter Simulation Conference*. Piscataway, NJ: IEEE.

Robinson S. 2005. Discrete-event simulation: From the pioneers to the present, what next? *Journal of the Operational Research Society* 56(6): 619–629.

Singh V. P. 2009. *System Modeling and Simulation*. New Delhi, India: New Age International.

Tocher K. D. 1963. *The Art of Simulation*. London: English Universities Press.

Tropp H. 1981. On doing contemporary history. In *History of Programming Languages*, ed. Wexelblatt R. L., pp. xxi–xxiii. New York, NY: Academic Press.

Weisberg D. E. 2008. *The Engineering Design Revolution*. http://www.cadhistory.net (free access).

Index

Milton Keynes UK
Ingram Content Group UK Ltd.
UKHW040052071024
449327UK00019B/512

9 780367 781385